Dairy Cattle and Milk Production
Prepared for the use of Agricultural College Students and Dairy Farmers

by Clarence H. Eckles

with an introduction by Jackson Chambers

This work contains material that was originally published in 1911.

This publication is within the Public Domain.

This edition is reprinted for educational purposes and in accordance with all applicable Federal Laws.

Introduction Copyright 2017 by Jackson Chambers

Self Reliance Books

Get more historic titles on animal and stock breeding, gardening and old fashioned skills by visiting us at:

http://selfreliancebooks.blogspot.com/

Introduction

I am pleased to present another title in the "Cattle" series.

The work is in the Public Domain and is re-printed here in accordance with Federal Laws.

As with all reprinted books of this age that are intended to perfectly reproduce the original edition, considerable pains and effort had to be undertaken to correct fading and sometimes outright damage to existing proofs of this title. At times, this task is quite monumental, requiring an almost total "rebuilding" of some pages from digital proofs of multiple copies. Despite this, imperfections still sometimes exist in the final proof and may detract from the visual appearance of the text.

I hope you enjoy reading this book as much as I enjoyed making it available to readers again.

Jackson Chambers

To

GEORGE LEWIS McKAY

FORMERLY PROFESSOR OF DAIRYING AT THE IOWA

STATE COLLEGE

MY TEACHER, MY COLLEAGUE

AND STEADFAST FRIEND

THIS BOOK IS DEDICATED

PREFACE

THIS book presents in printed form material that has been gathered by the author during the past ten years for presentation to students in the form of lectures. The author has had charge of a herd of from thirty to fifty cows for fifteen years. Among these have been many high-producing animals and all the leading dairy breeds.

It has been the aim of the author to bring together the essential information regarding the dairy cow in a compact and usable form. An immense amount of knowledge exists on this subject. It is found scattered through the publications of State Experiment Stations and of the Federal Government, in the Agricultural Press, and in the possession of practical herdsmen. An attempt has been made to bring this information together with the object of acquainting the student who expects to be a farmer with the principles he must understand and practice in order to be successful with dairy cattle. It is hoped it will be serviceable as well to the practical farmer interested in dairy cattle, who will find the material presented in such a way that it will assist him to care properly for his animals and to produce milk economically.

The author desires to express his appreciation of the assistance of his former student, Professor O. E. Reed, of the Kansas Agricultural College, who has supplied part of the photographs and assisted in preparing portions of the material.

C. H. ECKLES.

UNIVERSITY OF MISSOURI,
February, 1911.

TABLE OF CONTENTS

CHAPTER		PAGE
I.	Introduction	1
II.	Origin of Domesticated Cattle	9
III.	The Dairy Type	17
IV.	Holstein-Friesians	27
V.	The Channel Island Breeds	42
VI.	Ayrshires	63
VII.	Brown Swiss	74
VIII.	Minor Dairy Breeds — Dutch Belted, Polled Jersey, Kerry, French-Canadian	81
IX.	Dual-Purpose Cattle	87
X.	Starting a Dairy Herd	107
XI.	Selection of the Individual Cow	116
XII.	How Individual Selection is Made	132
XIII.	Selection of the Herd Bull	154
XIV.	Calf Raising	174
XV.	Calf Raising	191
XVI.	The Development of the Dairy Heifer	203
XVII.	Management of Dairy Cattle	210
XVIII.	Management of Dairy Cattle (*Continued*)	229
XIX.	Water and Salt Requirements	241
XX.	The Soiling System	248
XXI.	Feeding for Milk Production	254

TABLE OF CONTENTS

CHAPTER		PAGE
XXII.	Feeding for Milk Production (*Continued*)	274
XXIII.	Stables for Cows	294
XXIV.	Handling Manure; Material for Bedding	316
XXV.	Common Ailments of Cattle	324

LIST OF ILLUSTRATIONS

FIGURE		PAGE
1.	Group of Dairy Cows exhibited by University of Missouri at the Missouri State Fair	*Frontispiece*
2.	Skull of *Bos sondaicus* (Keller)	10
3.	Skull of *Bos primigenius* (Keller)	11
4.	Cross-section of a High-class Jersey Cow	19
5.	Cross-section of a High-class Fat Steer ready for Market	20
6.	Pure-bred Jersey Cow	*facing* 21
7.	Pure-bred Jersey Cow	" 21
8.	Pure-bred Ayrshire Cow	" 21
9.	A Pure-bred Jersey Cow of Great Capacity	" 23
10.	Examples of Well-developed Milk Veins	" 24
11.	Examples of Well-formed Udders	" 24
12.	Defective Udders	" 25
13.	Diagram showing Points of the Cow	" 26
14.	Pure-bred Holstein Cow	" 30
15.	Holstein Bull	" 32
16.	Pure-bred Jersey Cow	" 46
17.	Group of American Type Jersey Cows bred by the University of Missouri	*facing* 48
18.	Pure-bred Jersey Bull	" 49
19.	Pure-bred Guernsey Cow, "Dolly Dimple"	" 57
20.	Guernsey Bull, "Imported King of the May"	" 59
21.	Ring of Aged Ayrshire Cows at Alaska, Yukon Exposition	" 66
22.	Ayrshire Bull. A Famous Prize Winner	" 67
23.	Brown Swiss Cow, "Onetta"	" 76
24.	Group of Dutch Belted Cattle	" 81
25.	Pure-bred Shorthorn Cow, "Lula"	" 94
26.	Pure-bred Shorthorn Cow, "Lady Stratford"	" 94
27.	Red Polled Cow	" 100
28.	Jersey Cow, "Pedro's Ramaposa"	" 126
29.	"Bessie Bates," a Pure-bred Jersey	" 132

LIST OF ILLUSTRATIONS

FIGURE		PAGE
30. An Example of the Difficulty in selecting by Type	*facing*	132
31. Scales for weighing Milk		136
32. Milk Sheet		138
33. Form for keeping Milk Records, weighing Three Days per Month		139
34. Form for a Permanent Record		143
35. Frame for holding Milk Record Sheets		145
36. Bull Shed built at Purdue University	*facing*	171
37. Skim Milk Calves in Pasture	"	196
38. Calves tied in Stanchions for Feeding	"	196
39. Influence of Feed upon Size and Conformation of Dairy Heifers	*facing*	206
40. Influence of Age at First Calving upon Size of Dairy Heifers	"	208
41. The Influence of the Age of Calving and Feeding when Young upon Dairy Cows. — Light Fed Early Calving compared with Heavy Fed Late Calving	*facing*	209
42. Devices for marking Cattle		213
43. Correct Position of the Hands when Milking (Grotenfelt)		220
44. Instruments for Treatment of Udder Troubles		225
45. Apparatus for treating Milk Fever		237
46. Improvised Apparatus for treating Milk Fever		238
47. Plan of a Two-story or Loft Barn. (U. S. Dept. Agric.)		296
48. A One-story or Shed Type Barn	*facing*	297
49. A Good Example of a Round Barn	"	298
50. Interior Arrangement of a Good Barn	"	298
51. Arrangement and Plan of a One-story Barn. (U. S. Dept. Agric.)		301
52. Cross Section of a One-story Barn (Erf.)		303
53. Cross Sections (one half) of Barns showing Common Plans of Construction		307
54. Cross Sections showing Various Types of constructing Mangers		309
55. Stall constructed of Iron Piping		311
56. The King System of Ventilation		313
57. Trocar used for Bloat		337

DAIRY CATTLE AND MILK PRODUCTION

DAIRY CATTLE AND MILK PRODUCTION

CHAPTER I

INTRODUCTION

IMPORTANCE OF DAIRY FARMING — ADVANTAGES — ITS PLACE IN A SYSTEM OF PERMANENT AGRICULTURE

MILK, with its products, serves as one of the most important sources of food for all highly civilized nations. A large proportion of the best agricultural lands of the world are utilized for its production. Although milk and products of milk have been used to some extent for food as far back as history records, the general use of milk as food has come about only with the development of highly civilized nations. Martiny[1] points out that the native races of America, Africa, and Australia, which have never developed past the stage of barbarism, do not use milk as food. The primitive races of Western Asia and of Europe made use of milk, as have their descendants, and according to this author, to this fact is due in no small degree the great intellectual development of Europe and America.

To what extent this is true may be a question, but it is a well-known fact that the most prosperous agricultural nations and communities to-day are those in which the dairy cow is the foundation of agriculture. We have only to com-

[1] Benno Martiny, *Kirne und Girbe*.

pare Russia with Denmark, and Spain with Holland, to show what the dairy cow will do for a nation. If a list were prepared of our own states, selecting those where on the average the soil fertility is best conserved, the most intelligent system of farming followed, and the highest grade of intelligence found among the people, it would be a list of the leading dairy states.

The dairy cows of the United States number nearly twenty million, and the annual value of their products reaches the enormous sum of nearly one billion dollars. Only the corn crop and animals sold for meat exceed dairy products as a source of income to the American farmer. The rapid growth in the population of our country, together with a slow but constant increase in the per capita consumption of dairy products, makes it certain that the dairy cow will in the future occupy a still more important position. Some of the fundamental reasons why the cow is certain to play an important part in the future agriculture of America are pointed out in the following pages.

Relation to Fertility of the Soil. — It is now conceded that the conservation of the fertility of the soil is the greatest problem of agriculture. There is some difference of opinion as to the possibility of maintaining fertility where grain crops are sold from the farm. It is certain that whether it be possible or not, it is seldom done. So far in our history grain selling has meant selling fertility that has been stored up in the past ages, and has been followed by impoverished soils and unprofitable agriculture. On the other hand, we find farms in almost every locality, and even entire countries can be pointed out where the fertility of the soil has been vastly increased by live-stock farming. The most marked examples of this are in connection with dairy farming. The following table gives the fertilizing constituents of common feed stuffs and of dairy products. The value is calculated on the basis of nitrogen at 20 cents per pound, and phosphoric acid and potash at 6 cents per pound, which values are in use at present by chemists connected with the inspection of commercial fertilizers.

FERTILIZING CONSTITUENTS AND VALUE AS FERTILIZERS IN 100 POUNDS

	NITROGEN	PHOSPHORIC ACID	POTASSIUM OXIDE	VALUE PER TON AS FERTILIZER
Corn Fodder (with ears)	1.76	.54	.89	$8.76
Mixed Hay	1.41	.27	1.55	7.82
Timothy Hay	1.26	.53	.90	5.78
Red Clover Hay	2.07	.38	2.20	11.38
Alfalfa Hay	2.19	.51	1.68	11.39
Cowpea Hay	1.95	.52	1.47	10.19
Wheat Straw	.59	.12	.51	3.12
Corn	1.82	.70	.40	8.60
Wheat	2.36	.79	.50	9.59
Oats	2.06	.82	.62	9.97
Bran	2.67	2.89	1.61	16.08
Gluten Meal	5.03	.33	.05	20.58
Cottonseed Meal	6.64	2.68	1.79	31.92
Linseed Meal	5.78	1.83	1.39	26.90
Milk	.53	.19	.18	2.56
Cheese	4.52			18.08
Butter	.16			.64

It will be noted that in proportion to their market value dairy products take but little fertility from the farm. Wheat at $1 per bushel is worth $33.32 per ton, while it carries with it elements of fertility worth $9.59. Corn at 60 cents per bushel is worth $21.40 per ton, while it removes fertility worth $8.60. A ton of milk at $1.50 per hundred is worth $30 per ton and is worth $2.56 on the basis of the fertility contained.

The comparison is still more striking when cream or butter is sold. Since butter fat contains only carbon, hydrogen, and oxygen, it has no value as a fertilizer. The only element of fertility in butter is the small amount of nitrogen contained in the curd, amounting in value to only 64 cents per ton while the market value of this amount of butter at 30 cents per pound is $600.

A dairy cow weighing 1000 pounds voids about 12 tons of solid and liquid manure in a year, worth on the basis of the elements of fertility contained $30 in round figures. The Minnesota and Ohio experiment stations found from field experiments that barnyard manure has an actual value of from $2.50 to $3.50 per ton when applied to the land, depending upon the fertility of the soil. Under fairly good conditions at least 80 per cent of the fertilizing constituents of the manure may be returned to the soil.

But this does not tell all the story. The dairy farmer usually is a purchaser rather than a seller of grain, and by this means adds constantly to the fertility of his farm. The purchase of concentrated feeds rich in protein, as will be seen from the table, add a large amount of fertility to the farm. Furthermore, the keeping of dairy cattle usually means that a large proportion of the land is kept in grass, which makes it possible to prevent washing of the soil, which is responsible for the rapid deterioration of many farms.

It is a well-known fact that the yield of grain per acre of the agricultural lands of Denmark, Germany, and parts of England where dairy farming has been followed for a period of years has been materially increased. The Hosmer farm at Marshfield, Mo., for the past five years has yielded an average of 70 bushels of corn per acre on land that produced 15 bushels per acre 17 years ago, when the present owner established it as a dairy farm.

Adaptation of Dairying to High-Priced Land. — As a rule a thinly settled region is not a dairy country. When land becomes high in price, and it is necessary to secure a correspondingly larger income, the dairy cow usually comes into use. Exceptions to this are level rich lands that may be used for grain growing for long periods without exhausting the available fertility. Dairy husbandry is intensive farming, and a comparatively small area is sufficient to carry on such a system of farming. An example of what is possible along this line is a farm in Pennsylvania [1] of 17 acres from which milk to the amount of $2400 per year was sold, and young stock to the amount of $500. The purchased feed amounted to $625 per year.

Land on the Isle of Jersey, the annual rental of which is $50 to $60 per acre, is used for keeping the Jersey cow. Land in Holland, worth from $1000 to $2000 per acre, is used almost exclusively for dairy purposes. The same is true of most of the high-priced land in other parts of Europe.

The Cow a Cheap Producer of Human Food. — Henry says: [2] " Not only is dairying the leading animal industry of our country at the present time, but so it must continue indefinitely, for the reason that the cow is a more economical producer of human food than is the ox or pig." The following table from data gathered by the Missouri Experiment Station

---

[1] Farmers' Bulletin No. 242, U. S. Dept. of Agriculture.
[2] *Feeds and Feeding*, p. 401.

illustrates this fact forcibly. The comparison is made of the milk produced by a Holstein cow in one year and the composition of the carcass of a fat steer weighing 1250 pounds.

	18,405 LB. MILK	STEER, WT. 1250 LB.
Proteids	552	172
Fat	618	333
Sugar	920	—
Ash	128	43
Total	2218	548

The total amount of dry matter in the milk was 2218 pounds, all of which is edible and digestible. The steer, with a live weight of 1250 pounds, contained 56 per cent of water in the carcass, leaving a total of 548 pounds of dry matter. In this dry matter of the steer is included hair and hide, bones and tendons, organs of digestion and respiration; in fact, the entire animal, a considerable portion of which is not edible. The analysis of the steer's carcass was made from samples taken after grinding up together one half of the complete carcass.

The cow produced proteids sufficient for more than three steers; nearly fat enough for two, ash enough to build the skeleton for three, and in addition produced 920 pounds of milk sugar, worth as much per pound for food as ordinary sugar.

In the above comparison the cow was far above the ordinary, and for this reason the following additional data is given from the Missouri Experiment Station, representing the total constituents in the milk of several cows of ordinary dairy capacity: —

PRODUCTION FOR ONE YEAR

	BREED						
	Jerseys			Ayrshires		Holsteins	
Milk	8522	6775	6033	6276	6382	8685	8815
Protein	339	278	264	195	213	261	283
Fat	469	373	368	220	246	281	283
Sugar	393	290	254	305	317	437	375
Ash	63	51	45	40	39	56	62
Total Solids .	1264	992	931	760	815	1035	993

The above table shows that these ordinary cows all produced more protein in a year than was contained in the carcass of the 1250-pound steer. Three of them also produced more fat. The solids of all except two contained more ash than was found in the carcass of the steer. In addition the cows produced from 290 to 437 pounds of sugar each. The seven cows, representing three breeds, in one year averaged 970 pounds of total solids each, or nearly as much as was contained in the carcass of two steers.

A comparison of the feed consumed by the steer and the cows would be still more striking, since the steer required nearly two years of liberal feeding to build this carcass while the product from the cows was made in less than one year.

Constant Returns. — One of the advantages of dairy farming that appeals to the farmer with limited capital is the certainty of the returns. There is little of the element of speculation in this line of farming. The returns are not large at any one time, but steady throughout the year, and may be depended upon. The market price of dairy products varies on the whole less than almost any other class of farm products, making it safe for the farmer of small capital as well as for the larger.

The Labor Question. — The problem of securing sufficient and satisfactory labor is generally counted the greatest difficulty experienced in conducting a dairy farm. This difficulty arises from the necessity of treating the cow carefully at all times, and especially from the fact that the work becomes somewhat monotonous from having to be done regularly every day. While the labor problem is a serious one, it is no worse than experienced in conducting almost any other line of farming, and in fact under proper conditions may be less. The grain farmer crowds his work into a few months and requires a large amount of help for a few days or weeks only, and finds it almost impossible to secure, since he has no work to offer the remainder of the year. Work on the dairy farm is distributed throughout the year, and arrangements may be made accordingly. The special objections raised to the labor on the dairy farm are the long hours, the steady, regular work, and the nature of the work. To reduce the labor problem to the minimum, first of all the hours must be made as reasonable as in any other kind of farming. Provision should also be made for regular time off by each laborer in turn. The objections made to the nature of the work comes almost entirely from the conditions under which the work is done, and that may be removed. If the cows are milked in a clean, well-lighted, comfortable stable at reasonable hours, and modern methods of handling the manure and feed by overhead carriers are installed, the objections to the work will mostly disappear. In most localities by furnishing a comfortable house, a man with a family may be employed by the year with the best satisfaction to the employer.

CHAPTER II

ORIGIN OF DOMESTICATED CATTLE

CLASSIFICATION OF BREEDS

Origin of Domesticated Cattle. — There are no cattle native to America. All those found in North and South America are descended from animals brought from Europe. The domesticated cattle of Europe are descended from wild forms that formerly lived in Europe and Asia. Where and by whom cattle were first domesticated is unknown, as it took place in prehistoric times. Within recent years considerable light has been thrown on the subject by extensive investigations which have been made regarding the early types of cattle and their relationship to the domesticated breeds of the present.

This study has been carried on largely by comparing the skeletons of different breeds and types of cattle from all over the world. Other sources contributing to the knowledge of this subject have been extensive studies of bones found in ancient human dwelling places, as those of the Lake Dwellers of Switzerland. Ancient historical records and works of art which depict cattle have also been carefully examined by those studying this subject. The material that has been gathered is so fragmentary that even those who have given the subject most study do not agree on more than the general details. It is known from fossil remains that the ox existed in Europe before the glacial period, but it is uncertain whether the domestic cattle are descended from this form or from an Asiatic type. Among the most recent investigators on this subject is Kellar.[1]

According to this author the investigations so far indicate that the cattle of Europe are descended from two original types or species. One is called *Bos sondaicus*, the other *Bos primigenius*. This author believes cattle were domesticated long before the records of history began, while the ancestors of the present Europeans still dwelt in Asia. The *Bos sondaicus*, which was the first type domesticated, is still represented in Asia by the Banteng or native wild ox, found in small numbers on certain islands of the East Indies. Similar forms are said to be also found in a state of domestication in the same countries.

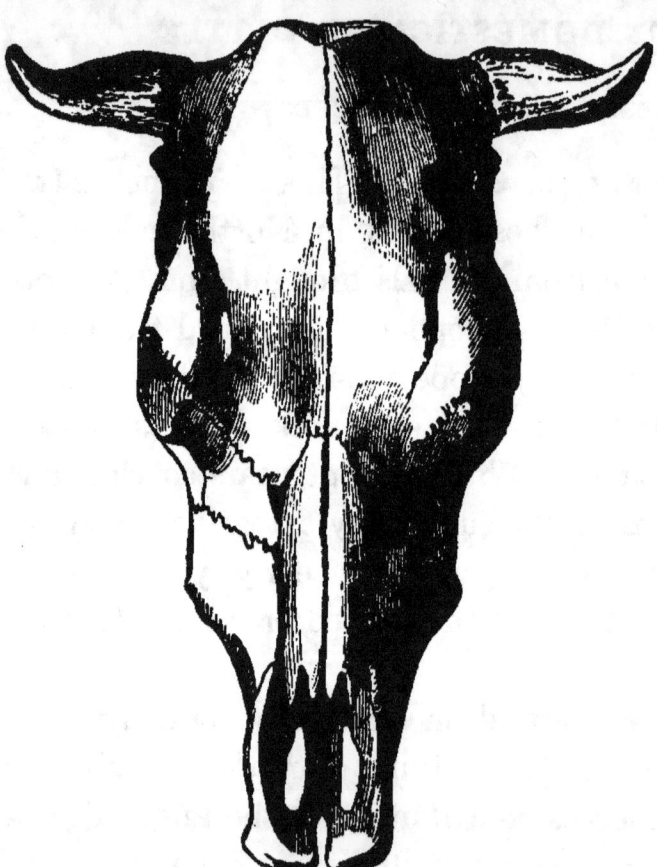

Fig. 2.—Skull of *Bos sondaicus* (Keller), showing the broad head and short horns.

These cattle were taken to Europe during the great migra-

[1] *Naturgeschichte der Haustiere.*

tions that took place, and were spread over the greater part of that continent. Numerous remains of this type are found in the oldest ruins of the Lake Dwellers in Switzerland. These cattle at this time were small in size, short-bodied, and had small horns. From this type have descended most of our

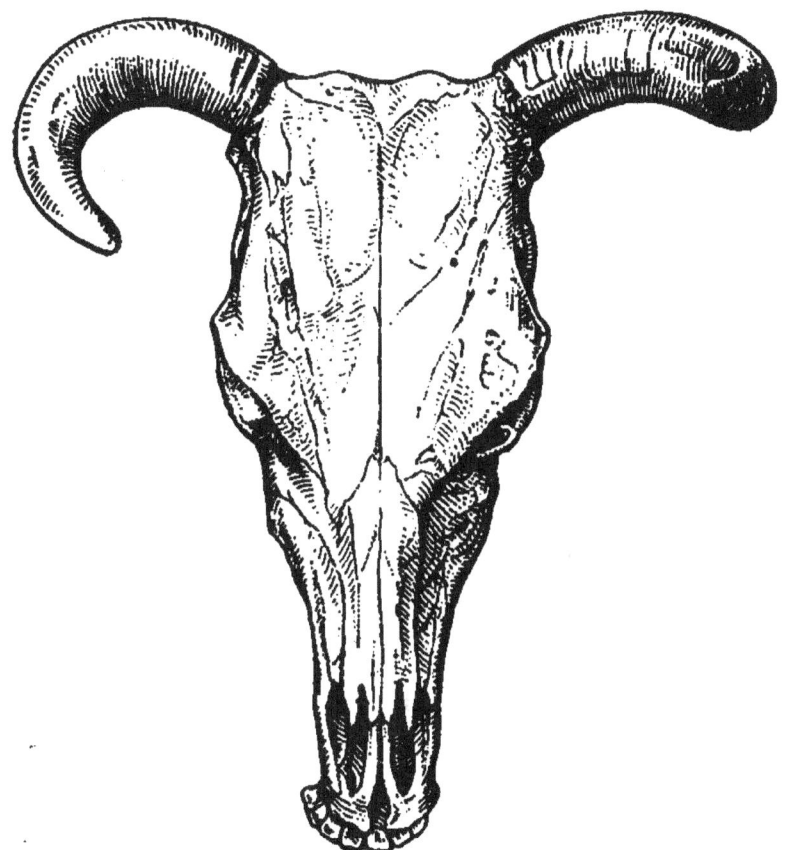

FIG. 3. — Skull of *Bos primigenius* (Keller), showing the long, narrow head and long horns.

breeds in use to-day, as the Brown Swiss, Jersey, Guernsey, and all of the breeds of England except the Longhorn and Scotch Highland. However, part of the English breeds, especially the Shorthorn and Ayrshire, while having this type as a foundation were mixed in the early days of these breeds with the blood of the *Bos primigenius* type through crossing with Holland cattle.

The *Bos primigenius* was an immense, powerful animal with tremendous horns. Historical records show this form existed in a wild state in Europe until the twelfth or thirteenth, and possibly the fourteenth century. This animal was apparently domesticated in Europe within historic times. From it are descended the cattle of Holland and other parts of North Europe, the large, long-horned cattle of Hungary and adjacent regions, the Fleckvieh or Spotted Cattle of Switzerland, and the Longhorns and Scotch Highland breeds in England.

The chief basis of classification is the skull, which is quite different in the two types. In the Holland or Holstein breed, for example, we have the long, narrow head, indicating descent from the *Bos primigenius*, while in the Channel Islands breeds we find the head broad and short, which is characteristic of those breeds descended from the *Bos sondaicus*.

Origin of Breeds. — Varieties or breeds of cattle came into existence first of all as a result of environment, such as climate, food, and the nature of the surroundings. In the early times, with no organized means of transportation, there naturally was little exchange of animals from one locality to another, and probably little, if any, attempt at improvement. The effects of natural conditions were allowed to work out almost undisturbed by the agency of man. Breeds formed by such means may be called natural breeds.

On the continent of Europe the breeds and sub-breeds are almost innumerable, and they are mostly breeds originated in the manner mentioned. In Great Britain alone ten or twelve distinct breeds have been originated. Up to about the middle of the eighteenth century these natural influences were the chief factors developing distinct breeds. About this time a great interest was aroused in England in regard to improving the quality of the cattle and other domestic animals of Great Britain. This exceedingly important movement, which was largely the result of the work of Robert Blakewell, spread more or less to other cattle-breeding countries. The beginning of modern improved breeds is to be traced to this great movement.

The methods used were careful selection of breeding animals, liberal feeding, and general good management. In most cases, as, for example, in improving the Shorthorns and Ayrshires, crossing and inbreeding was at first practiced. At the present time the efforts of cattle breeders are directed toward further improvement in the breeds already in existence, and not toward the establishment of new breeds, because it is generally recognized that selection may be made among those already established to suit any conditions under which cattle may be profitably kept.

Value of Breeds. — The breed is only one of many factors to be considered in carrying on profitable milk production. In some cases the value of the breed is overestimated, but more often the reverse is true. Our present dairy breeds represent the efforts toward improvement in certain definite lines made by several generations of breeders. It would be folly for a man to attempt to start at the beginning to build up for himself what it has taken a century or more to build by others. By making use of animals of a highly developed breed adapted to the purpose for which they are to be used, he is taking advantage of all the work that has been done, and is starting in at the highest point of advancement reached by other breeders.

On the Market. — Cows of a distinct dairy breed usually, and rightly, sell for more than the same number of cows of mixed or unimproved breeding, even if the latter are known to be equally good as dairy cows. The cows of a distinct dairy breed are worth more to the buyer, because he can reasonably expect these animals to show the typical character of the breed to which they belong in production of milk, in disposition, and in other breed characters. Further, he can reasonably expect that these cows, when mated with a male of the same line of breeding, will produce offspring having the same typical breed characters. A cow of mixed breeding, even if a good dairy cow, or an unusually good milker in a breed where milking qualities are not generally found, cannot be counted upon to reproduce herself in her offspring. It is a well-known fact in animal breeding that the longer a certain character has existed in a breed, the more certain it is to be transmitted.

Pure breeds have been bred generation after generation with certain objects in view, and in course of time these characters become fixed as breed characters, and are transmitted. It is easy to understand why the chances are good for getting a good dairy cow if the ancestors are Holstein, known to have been bred about 2000 years in one locality and noted for hundreds of years as great dairy animals, or if the parents are Jerseys bred for 500 years, or longer, along one line.

Classification of Cattle. — No system of classification has yet been devised that can be applied in more than a general way to the individuals that make up the great mass of cattle. If we undertake to arrange them by breeds, we find, in addition to the numerous pure breeds, animals with all possible mixtures of the blood of two or more breeds, or with more or less improved blood mixed with the scrub or unimproved. If we should attempt to arrange them according to the purpose for which they are adapted or kept, we would have a constant gradation from the extreme of beef to the extreme of dairy development.

It is even difficult to arrange a suitable classification of the pure breeds, since the animals may vary greatly within a breed due to environment and treatment. The following descriptive terms are in common use : —

Unimproved, Scrub, or Natives. — These terms generally indicate that the animal does not carry more than at least a small amount of the blood of any of the improved breeds. Typical scrubs are not numerous except in those sections where very little attention is given to cattle raising. The term " scrub " is often applied also to inferior animals of any breeding.

Cross-bred is a term used to indicate that the animal is the offspring of parents of distinct breeds, either high grades or pure bred.

Grade. — This term is generally used with a certain breed name, as Grade Jersey or Grade Shorthorn. This means that the animal in question has one half or usually more of the blood of the breed mentioned. When the proportion of the pure blood is large, the animal is called a " high grade." The proportion of the blood predominating may be so great that for all practical purposes the animal is the same as a pure bred, but it cannot be called a pure bred no matter how many crosses have been made, and such animals cannot be registered in the various Herd Books.

Pure Bred. — The term "thoroughbred" is often improperly used instead of the proper term, "pure bred." The term thoroughbred is properly applied only to the well-known English breed of horses. Pure-bred cattle, as understood in America, are those whose ancestors came from the native home of the breed in question and conformed to the requirements of this breed here. This blood must be kept pure and unmixed, and records must be available showing the descent from these ancestors. The records of descent of these animals are kept in a systematic manner by associations formed for the purpose by those interested.

The breeds of cattle common in America are usually classified as dairy, dual-purpose, and beef.

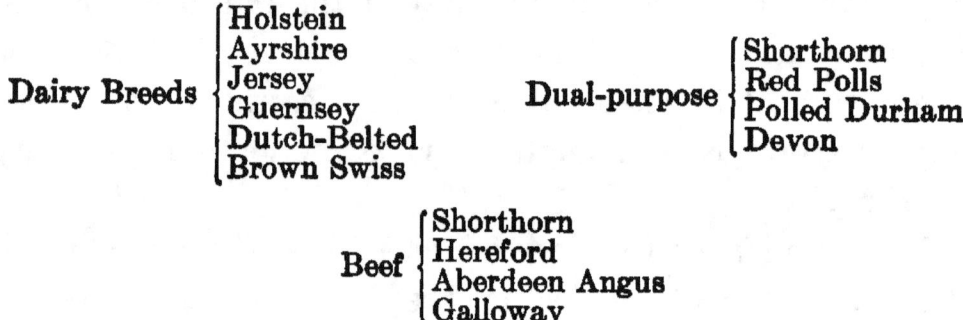

In addition to the above, small numbers of French-Canadian, Kerry, and Polled Jersey cattle, all to be classed as dairy breeds, are found in certain localities in America.

CHAPTER III

THE DAIRY TYPE

THERE is but one entirely satisfactory way to select cows for dairy purposes, and that is by records of the production of each individual, made by the use of the scales and Babcock test. Since up to the present time individual milk and fat records have been kept for only a small proportion of the cows used for dairy purposes, the selection of individual cows on this basis is impossible in more than isolated cases. Most selection must be based upon conformation, or the degree to which the animal approaches what is known as the dairy form or type. The excessive development of the function of milk production through generations of selection and breeding in that direction has brought about certain characteristics in the conformation of the animal that may be taken into account in judging of the development of these functions.

The breeders of Jersey Island in 1834 formulated the first scale of points for dairy cattle. At the present time the breeders' associations have prepared for each breed a carefully drawn scale of points that are of assistance in acquiring a skill in the selection of cows by conformation. A scale of points undertakes to describe the conformation of the animal that in the judgment of the author denotes the highest development of the characteristics sought. The comparative importance of the parts described is represented by points that total 100 for a perfect animal. The lack up to the present of a real scientific basis for preparing a scale of points makes them unsatisfactory in many ways, but of great general value, especially to the beginner.

The General Characteristics of the Dairy Type. — A person familiar with cattle in general, but not with highly developed dairy cattle, looking for the first time upon a high-class dairy cow in full flow of milk would have his attention especially directed to three points, as follows: —

1. The extreme angular form, carrying no surplus flesh, but showing evidence of liberal feeding in her vigorous physical condition.

2. The extraordinary development of the udder and milk veins.

3. The marked development of the barrel in proportion to the size of the animal.

These three statements should be kept in mind as describing the special characteristics of the dairy animal as compared with those bred for beef, or with inferior dairy animals. Sometimes the error is made of attributing this lack of flesh, so characteristic of a good dairy cow, to insufficient feeding. The dairy cow does not, however, have the same appearance as an animal not of the dairy type that is thin in flesh on account of insufficient feed. A high-class dairy cow never carries much flesh when in full flow of milk. The stimulation to produce milk is so strong that all the feed she can consume and digest is utilized in producing milk. Such an animal, although thin in flesh, has an alert, vigorous appearance, her hair is soft and healthy, the skin pliable and loose, her paunch is full, and a general appearance of thrift and contentment is noticeable. An animal thin in flesh on account of insufficient feed has a stupid appearance, and shows a lack of vigor, while the rough, long hair stands on end. The paunch may be large or not, depending upon the bulkiness of the feed consumed by the animal.

The Dairy Form. — So characteristic is this angular appearance of the dairy cow that an animal that does not show this form when in full flow of milk should not be selected. It

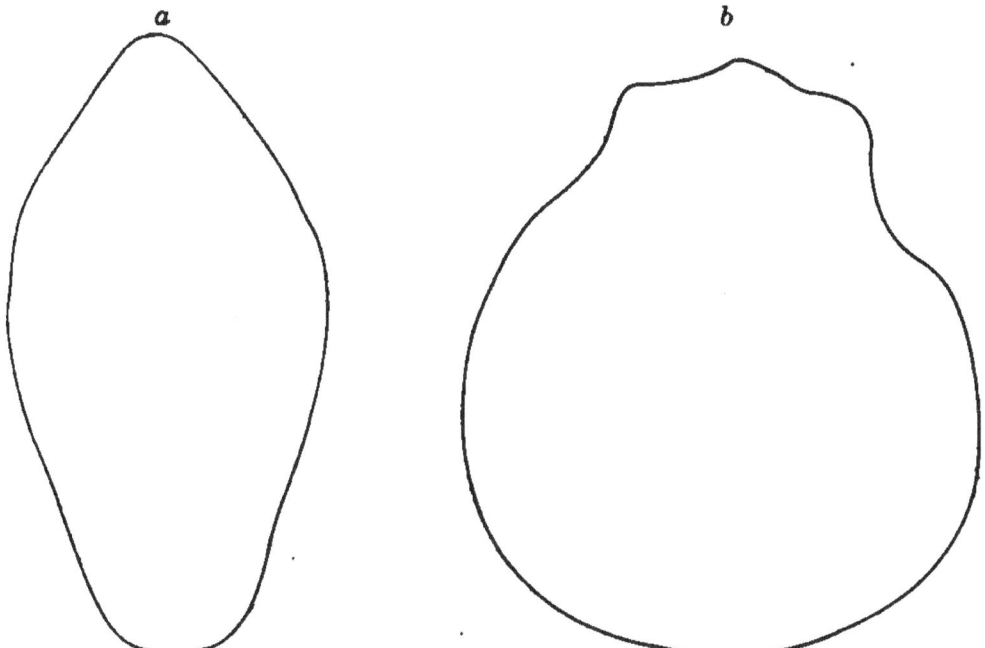

Fig. 4. — Cross-section of a high-class Jersey cow. *a*, at heart girth; *b*, at paunch. Weight, 900 pounds.

should be understood that it is natural for a cow to fatten considerably towards the end of her milking period and when dry. This surplus fat is mostly taken from the body during the first three or four weeks after calving. It is impossible with any ration to fatten a high-class dairy cow during the best part of her milking period, or even to keep the fat on her body at calving time from being removed, during the first few weeks she is in milk.

The cow that shows these characteristics to a marked degree

is said to have a good dairy temperament. This means she is endowed by nature with a strong stimulation to produce milk, and uses practically all the nutrients she can digest for milk production. This accounts for the spare form and absence of any surplus fat, even when the animal evidently has abundant food. As a result of the above, a high-class producing cow when in milk is usually thin and sharp over the withers,

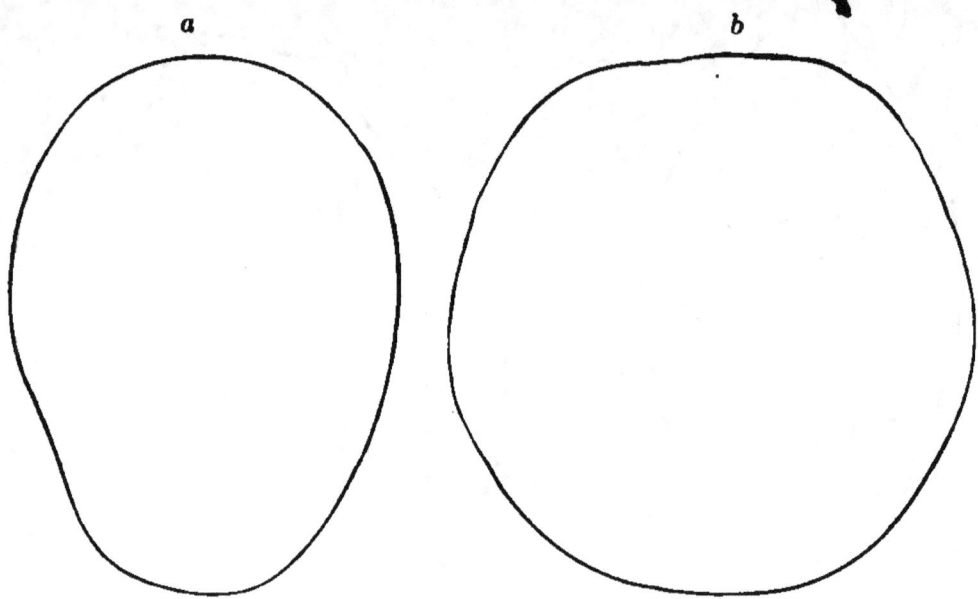

Fig. 5. — Cross-section of a high-class fat steer ready for market. *a*, heart girth; *b*, paunch. Weight, 1500 pounds.

her backbone stands out strong and prominent, her hips and pelvic region stand out almost free from flesh.

Figs. 4 and 5 show the contrast between highly developed beef and dairy animals. These cross-sections were made by the method devised by the Missouri Experiment Station.[1] When the cow is dry or nearly so she should carry more flesh than when in full flow of milk, and should not be criticised on this account. The breed type should be taken into account as well, and the mistake avoided of judging all by the same arbitrary standard.

[1] Waters, Proceedings Society for the Promotion of Agricultural Science, 1908, p. 71.

Types of Cows. — Fig. 6 is a good illustration of a cow lacking in dairy temperament, although a pure-bred animal of one of the leading dairy breeds. This animal has a good barrel, and a splendid digestion, an unusually good heart girth, and good skin and hair. She lacks the stimulation necessary to use her food for producing milk. This is shown by her thick withers, thin covering of flesh over the back, and general smooth beefy appearance. Fig. 5 illustrates the other extreme. This cow has the tendency to produce milk so strongly developed that she uses all the food she can eat and digest for this purpose and carries no surplus flesh. Her withers are thin and sharp, her back and pelvic region angular and bony, although she receives a liberal ration.

A cow should also be expected to carry somewhat more than her normal flesh for a short time after calving, but this beefy appearance should disappear within a month or less.

Limitations of Selection by Type. — The selection of dairy cows by type as indicated is often uncertain. Still the practical breeder or dairyman must select most of his animals in this way. The limitations should be understood. Any one familiar with dairy type will seldom fail to choose between a high-class animal and an inferior one, as, for example, between the cow shown in Fig. 8 and that in Fig. 7. It is usually easy to choose between a cow producing 350 pounds of butter fat in a year and one producing 150 pounds. However, as between the good and the extraordinary cow, type gives little upon which to base selection.

The author has yet to see a cow of extraordinary dairy quality that does not conform to the descriptions given in the following pages. While in some cases these cows would fail to score high on account of not conforming to the score card in some respects; for example, on account of a weak fore udder or a sloping rump, they showed the important characteristics described later in every case. On the other hand, some cows exhibit all the characteristics of high producers in regard to type, but fail in not having sufficient stimulation to produce large quantities of milk. To judge a dairy cow with any accuracy, she must be in milk, and preferably near the best stage of her milking period. A dry cow offers in most cases very little upon which to base judgment. A dairy cow thin in flesh from underfeeding is also in a condition that makes it almost impossible to form any estimate of her value.

Development of the Barrel.—The dairy cow that is a heavy producer must have large organs of digestion in order to utilize the enormous quantities of feed necessary to produce large quantities of milk. This results in the development of a large barrel, as that part of the animal's body between the fore and rear legs is called. A high producing cow has wide-sprung ribs and a deep abdomen, giving great capacity for the digestive tract and other vital organs. An animal lacking in this barrel capacity cannot use sufficient feed to be a large producer. The age of the animal has some influence on the size and depth of the abdomen. The depth of the barrel increases some with the age of the cow. The feeding of a ration consisting mostly of bulky feeds, as hay and silage, also tends to give the appearance of a greater barrel capacity from the greater contents of the digestive tract. In considering the barrel development of a cow, the depth as viewed from the side should be observed, then the width as viewed from behind. Some animals show a great depth, but on account of being narrow have no more real capacity than another animal with less depth, but greater width.

Fig. 9 is an illustration of a Jersey cow of great capacity. She shows an exceptionally good development of the barrel. Fig. 7 is a Jersey equally as well bred as the former, but on the other extreme in dairy capacity. She shows an unusually small development of the barrel, in keeping with her inferior dairy qualities.

Circulation. — After the food is digested and absorbed into the circulation it must be carried to other organs of the body, and undergo many changes before it is secreted in the form of milk. A strong, active circulation is of great importance, since without it the whole organism lacks tone. A large heart girth is usually assumed to indicate a large capacity of the heart and lungs. There is, however, some question as to the relation between the size of the body at the point called the heart girth and the size of the vital organs, but there are but few exceptions to the rule that cows of great milk-producing capacity, especially those that continue to produce for a series of years, have more than an average development in this respect. A soft, pliable skin is also an indication of a good circulation. When the animal has good " handling qualities," it means the small blood vessels below the skin are active and that the animal is in good health. A clear bright eye is also an index of a good circulation.

The Milk Veins and Milk Wells. — The most important point to be observed regarding the circulation is the development of the milk veins. The blood, after supplying the udder with material for milk secretion, starts back towards the heart through the milk veins. One of these opens on either side near the front line of the udder attachment to the body, and passes forward just beneath the skin. These veins crook back and forth more or less, in some cases divide into two or more divisions, and finally pass upwards through one or more openings in the wall of the abdomen into the body cavity. The portion of the veins from the udder to the opening through which it passes into the abdomen is called the milk vein. The opening in the abdomen through which the vein passes is popularly known as the milk well. Fig. 10 shows exceptionally good development in this respect. The milk vein is one of the most reliable indications of dairy capacity, since a large production of milk calls for a large quantity of blood to pass through the udder, and a large milk vein denotes such a circulation. The size of the milk vein is influenced to a great extent by the age of the cow. In a young animal the vein is smaller and more elastic than in the aged cow. When a cow is producing the maximum amount of milk, the veins are larger than is the case when the same animal is dry. The milk wells, on the other hand, remain of practically a constant size after the cow is once mature. In judging a dry cow, or one far advanced in the period of lactation, the size of the milk well is of greater importance than the size of the milk veins.

The Udder. — The development of the udder is of the greatest importance in selecting the cow, especially in regard to its size and shape. In the manufacture of milk, the food of the cow is first digested and becomes blood, then passes through the circulation to the udder. Since this gland is responsible for the secreting of milk from the blood, its size and development are of the greatest importance of all as indicating the dairy qualities of the cow.

It is not the size of the udder alone that is important,

but the number of active secreting cells. An udder gland filled with inert cells and fatty tissue is not effective. This is illustrated by *b* in Fig. 12. This is a large, well-formed udder, but the cow is a very moderate milker. Her udder is nearly as large after milking as before. The best type of udder has an especially long attachment to the body, both in front and behind. A good circulation of blood and healthy tissue is indicated by the soft, pliable skin and prominent veins. Before milking the udder is naturally considerably extended; after milking it should be greatly reduced in size and show an abundance of loose skin and a soft, pliable texture. Fig. 11 shows splendid examples of well-balanced udders.

The attachment to the body in the rear should extend well up behind. Special attention should be given the fore udder, as this part of the gland is especially subject to incomplete development. Fig. 12 illustrates defective udders. No. A in this illustration, in spite of her weak fore quarters, has a large udder capacity, and is a heavy milk producer. The quarters should be even in size, without deep indentations between. The teats should be of proper size for convenient milking, and evenly placed.

For show purposes especially, the shape and symmetry of the udder is especially important. From the standpoint of production the essential thing is to have sufficient udder capacity to admit of the secretion of a large amount of milk, with teats of such size as to admit of convenient milking. When a cow is dry it is impossible to judge accurately of the development of her udder. However, a large number of loose folds of skin, showing an abundance of room for expansion when the udder is filled, may be taken as an indication that the udder will develop in a satisfactory manner. The length of the attachment to the body should be especially noted in this condition. But little can be judged regarding the future size and shape of the udder in the calf or heifer until the time for calving approaches. The size and placing of the teats may be observed and judged with more accuracy than can the future development of the udder.

The Score Card. — Score cards as adopted by the associations concerned will be found for each breed with the other matter pertaining to that breed. From these a detailed study may be made of the type of that breed as described by those most concerned. Before using the score card the reader should be familiar with the points of the animal as illustrated in Fig. 13. The use of the score card is of advantage to the beginner as a means of impressing the points to be taken into account and their relative importance. The value of the score card decreases as experience is gained. Judging of cattle in the show ring is done entirely by comparison.

CHAPTER IV

HOLSTEIN-FRIESIANS

ORIGIN AND CHARACTERISTICS

Origin and Distribution in Europe. — This breed originated in Holland, and more especially in the province of Friesland. They are not, as the name would indicate, natives of the duchy of Holstein, which is a province in North Germany. The compound name Holstein-Friesian, the official name of this breed in America, resulted from a union of the Holstein Breeders' Association and the Dutch Friesian Association in 1885. In common usage now in America the breed is called Holstein.

The ancestor of this breed, according to Kellar, was the *Bos Primigenius*, the wild ox of Europe. This breed is one of the oldest in existence. Historical references indicate that they have been bred in the same region for at least 2000 years, and probably there has been very little, if any, mixing with outside blood. In the time of Cæsar the region now part of Holland was famous for its cattle. In the ninth century Holland was well known for its cheese and butter. According to Motley, in the seventeenth century Holland exported annually immense quantities of butter and cheese, and was noted for its immense oxen. This reputation has been maintained ever since, and during all these centuries cattle rearing has been almost the exclusive business of the Holland farmer. To-day no fruit is grown, and very little grain. The caring for cows, growing and preparing feed for them, and utilizing the milk for butter and cheese manufacture monopolizes the attention of the farmers.

This breed is best developed in its native home, the province of Friesland and across the Zuyder Zee in North Holland. This breed has been the parent stock of several others, which through local influences have been somewhat modified from the original. Most prominent among these are the Oldenburg, East Friesian, East Prussian-Holland, and the Flanders of Belgium.

The Holland cattle, or their descendants in these latter mentioned sub-breeds, are now distributed over a large portion of North Europe, extending into Russia. In the seventeenth and eighteenth centuries Holland cattle were taken to England, where, according to Professor Low, the eminent English authority, they were a factor in the formation of the Teeswater or Shorthorn breed, from which our modern improved Shorthorns are descended. It is also believed by good authorities that Holland blood was an important factor in the foundation stock from which the present Ayrshire breed is descended.

Conditions in Holland. — The best part of Holland is mostly below the level of the sea, which is kept out by enormous dykes. The land is very fertile, and almost entirely used for growing grass. The farmers, who are mostly tenants, pay from $30 per acre upwards annual rent. The land, which is seldom bought or sold, is valued at from $800 to $2000 per acre.

In no other part of the world does the cow receive such careful attention as in Holland. The cattle are placed in the stable, which is separated by a door from the living room of the family, about the first of October, and remain there as a rule constantly until about the first of May. The stables and the cows are kept in a condition beyond criticism from a sanitary standpoint. If an animal becomes soiled with manure, she is washed and cleaned carefully before milking. The feed is mostly hay, with a small allowance of linseed or other cake.

In the summer season the cattle are kept on the pastures and not brought to the barn. The milkers carry the milk from the pasture rather than fatigue the cows by driving them. If a cold wind blows up, the cows are at once blanketed in the pasture. Great care is taken in raising stock only from the best animals. Only a few bulls and about one fifth of the heifers are raised, and these from the best milkers only. The surplus calves are sold for veal, and the cows as a rule are sold for beef at an age of eight or nine years. In Friesland the milk is used largely for butter making and in North Holland mostly for cheese making.

Importations and Distribution in America. — A few were imported as early as 1795, but were not kept pure. The first importations that were kept pure were made into Massachusetts in 1861.[1] Only a few were brought over before 1875, but from this date until 1885 about 10,000 were imported. From these are descended most of the animals of this breed now in America. After 1885 none were imported until about 1903, partly on account of the high fee required for registering imported animals and on account of the prevalence of foot and mouth disease in Holland.

The first breed association, called the Holstein Herd Book

[1] Houghton, *Holstein-Friesian Cattle*, p. 17.

Association, was formed in 1873, and five years later the Dutch Friesian Association was organized. In 1885 the two united under the present name, Holstein-Friesian.

At the present time this breed ranks in number second only to the Jersey among the dairy breeds in America, and they are gaining in numbers and popularity very rapidly. At present they are found in every state, but the largest numbers are in the Eastern states, New York, Wisconsin, Pennsylvania, and Ohio having especially large numbers of this breed. Up to the present there have been registered something over 225,000 head, of which about one third are bulls.

Form and Characteristics. — The Holsteins are the largest of the dairy breeds. The average weight of the mature cow is 1200 pounds, but individuals vary from 1000 to 1600 pounds. The bulls weigh from 1800 to 2200, as a rule. The color markings are variegated black and white. As a rule the breeders have preferred animals on which the two colors are about evenly divided. The colors are always sharply defined and not blended. Fortunately this breed has never been injured by a color fad, although there is recently a tendency to favor those having more white than black.

As a breed the Holsteins have the best disposition or temperament of any dairy breed. In this respect they resemble the Shorthorns more than any other breed. While cows of this breed as a rule have plenty of nervous energy, which is necessary to high dairy production, they are not nervous in the common meaning of that term. Where Holsteins and other more excitable breeds are kept together, the contrast is easily noticed. A change of milkers, or any sudden disturbance, as the presence of a stranger or a dog, will produce little or no effect on most Holsteins, while cows of some other breeds will show a marked change in milk production. The Holstein is less alert and active than the other dairy breeds, but her nerves are well under control. This is of considerable advantage on account of the usual necessity of having dairy cattle handled by men more or less careless and inefficient.

In the descriptions found in the early volume of the Advanced Register of this breed, the cows are classified according to form as follows: —

Milk and Beef Form.

Milk Form.

Beef and Milk Form.

Beef Form.

This form of description is little used now. It was arranged for use in examining cows for admission to the Advanced Register by inspection. Cows are now admitted by an official test, and no description is required. The milk and beef form is the type of most of the imported cows, and the prevailing type. They have the wedge shape, but not especially pronounced; the shoulders are rather thick, deep, and broad. The barrel is round, with hips and loin broad and full. The quarters are straight and rather full.

The milk form is more angular in general appearance, the shoulders thinner and the wedge shape more pronounced, the loins and hips broad, thigh thin and incurving. The other two divisions of the classification are seldom used but indicate a more pronounced tendency towards flesh production.

The cattle of this breed, as found in Holland, on the average show somewhat more of a beef type and less pronounced dairy type than those found in America. The Holland farmer generally sells his cows for beef while still comparatively young, and also expects to derive considerable income from veal production. For these reasons he insists on some meat-making capacity. Those brought to America have generally been those having the best developed milking indications, since their reputation in America is based upon their milk-producing capacity.

The type of this breed as bred in America is believed by some to be changing slowly since they were first introduced. This change is attributed by those claiming to recognize such a change to the fact that as a rule the judges at fairs within recent years have had a strong prejudice in favor of the extreme dairy type, brought about by association with other breeds or as a result of agricultural college instruction. The winners in the show ring in recent years have usually been cows of medium or smaller size and those showing the most pronounced dairy form. In judging Holsteins care must be taken especially by the agricultural college student to keep the breed type in mind and not judge them according to the standard of other smaller breeds.

It is probably safe to say, the typical Holstein as now bred in America is slightly smaller than the Holland animal, and a larger proportion have a pronounced dairy type. It is also claimed by some breeders that the per cent of fat has been increased by American breeders. While it is possible that the strong efforts now being made in this direction by the leading breeders has resulted in richer milk from certain herds of selected animals, there is no evidence to show that the average of the breed has been changed. Data based upon seven-day official tests is of little if any value in this connection.

Dairy Characteristics. — The following summary shows the yield and composition of the milk produced by this breed as reported for animals owned by American experiment stations. Only yearly records are used which represent registered cows.

	AVERAGE	No. Cows Represented
Pounds milk per year . .	8699	83
Per cent fat	3.45	83
Yield of fat per year, lb. . .	300	83
Per cent total solids . . .	12.29	9

The fat represents 28 per cent of the total solids, as compared with 34.5 for the Jerseys.

The Holsteins produce more milk on the average and at a cheaper cost for 100 pounds than any other breed. The per cent of fat averages the lowest. The fat globules are small, rather variable in size, and do not show much yellow color. On account of the small globules, the cream does not separate so quickly nor so completely by gravity as is the case with larger fat globules. The lack of color in the fat results in the milk and cream showing much less color than if it was of equal quality but the product of a Jersey or Guernsey. The lack of color is of some disadvantage in selling market milk, since in the popular mind color erroneously is considered an index of the richness. The Holstein breed is well adapted for supplying milk for market. The small fat globules are an advantage, as they allow the milk to be handled easily without churning. The yield is large, and the fat and solids not fat well balanced for human food. If most of the cows in a heavy producing herd are fresh at the same time, the milk may be below standard in fat and solids at times.

This may be remedied by standardizing the milk. This consists in bringing it to a uniform per cent of fat by using a separator and taking out some of the skim milk if too low in fat, or taking out a portion of cream if above the standard. The milk of the Holstein cow is well adapted for calf raising, if it is desired to feed the calf whole milk. The solids not fat are high in proportion to the fat, and the latter is sufficient for the best results.

Excessively rich milk is not suitable for calf rearing, as has long been known by practical experience and recently confirmed by experimental work. As beef producers the breed ranks high for a dairy breed. As is the case with the other dairy breeds, the gains are made as rapidly and as cheaply as with animals of beef breeds. The market price is always lower than for animals of the beef breeds, partly as a result of prejudice, but mostly on account of the smaller proportion of high-priced cuts and the greater amount of offal.

The calves are especially well adapted for veal production. The average birth weight is 90 pounds, and they gain rapidly during the first few weeks. This breed is not especially early maturing. The heifers come into milk usually between twenty-four and thirty months. The breeding qualities of this breed are of a high order, being regular and sure breeders.

The Holstein cow is adapted for rather level, rich pastures, and where liberal feeding is practical. As grazers on hilly or scanty pastures the breed is surpassed by the Jersey and Ayrshire, especially the latter. They are heavy consumers of roughness.

The strong points of this breed are: the high average milk production; the marked vigor and strength of constitution; the strong vitality of the calves; the good breeding qualities;

the adaptation of the breed for veal production, and the quiet, contented disposition.

The low per cent of fat in the milk is generally considered the weakest point of this breed.

Families. — There are no well-defined families in this breed. The aversion of the Holstein breeders to in-and-in breeding has largely prevented the formation of families. Holstein breeders often refer to animals under what seems a family name; for example, referring to a certain animal as a Johanna or a DeKol. This means the animal in question is a descendant of the noted animal mentioned. As a rule, certain cows have been given prominence in Holstein affairs more than bulls. At the present time cows having large official records and their descendants are in greatest demand among breeders. Since a large number of official records have been recorded, the value of certain bulls as sires of high testing cows is shown clearly, and by this means certain bulls have more recently come into prominence and their descendants are especially sought after.

The Advanced Registry has been one of the important factors in the increase of popularity of this breed in America. The Holstein-Friesian Association, and especially Mr. S. Hoxie, of Yorkville, N.Y., should be credited with the introduction of this system, since adopted in somewhat different forms by other dairy cattle associations in America. The plan is to make record of dairy performance, in addition to regular registration.

The first entries were made in 1886. Under the original plan the animals were admitted on making certain milk and butter records as the result of tests made and reported by the owners themselves and after an inspector had scored and examined the animal.

This association was the first to adopt the Babcock test as their official method which was done in 1894. At first the plan was adopted of printing the records in the form of butter calculated on the basis of 80 per cent fat. Since this overrun is higher than is ordinarily secured in making butter, this method was criticized severely until, in 1903, the rules were changed. At the present the supervisor of the test reports the milk and fat yield only. If it is desired to express the result on the basis of estimated butter, one sixth should be added to the fat. In examining a Holstein pedigree it should be observed whether the records are given in butter fat or butter, and if the latter, upon what basis it is estimated. The majority of the Holstein breeders still use the 80 per cent basis in giving the records of butter production by their cows. The majority of the tests are for seven days, but they may be for any longer period. Breeders are encouraged to test again after eight months. The tests are conducted by the experiment station in the state where the cow is located. If the cow reaches certain requirements, she is given a number and entered in the Advanced Registry. She is afterwards commonly referred to as an A. R. O. cow.

The minimum requirements for admission are the following: —

    2-year-old . . . . . 7.2 pounds fat in 7 days
    3-year-old . . . . . 8.8 pounds fat in 7 days
    4-year-old . . . . . 10.4 pounds fat in 7 days
    5-year-old . . . . . 12.0 pounds fat in 7 days

Every day of increased age at date of calving increases the requirement .00439 pounds of fat until five years is reached. After a cow is entered in the Advanced Register on a seven-day test she is eligible to a year's test. The owner keeps the daily records of milk produced, and the product for two days each month is weighed and tested by the representative of the experiment station.

Objections to Seven-Day Tests. — The Holstein-Friesian Association should be given credit for starting the system of testing cows, and for being the first to adopt the Babcock test. The dairyman is interested in knowing what it is possible to get from a cow in a week, but much more interested in knowing what the cow in question will do in a year. The cow that makes a large weekly record is not always an animal that makes a large or even fair record for a year. There is an increasing demand for yearly records, and it is probable that in the near future all records will be made by the year and the short period will be dropped. A more serious objection to the seven-day test is that it makes it possible to obtain an abnormally high per cent of fat, as first shown by the writer.[1] The average per cent of fat for the breed is 3.45, but many seven-day tests are now reported with a per cent of fat over 4.50, and several above 5 per cent.

Many of the largest yearly milk records made by this breed were by some of the early imported cows or their immediate descendants. These records, however, were all private, and are not ranked with those authenticated by experiment stations. Some of the best of the early private milk records are the following: —

Clothilde	26,021 lb. milk 1 year
Clothilde 2d	23,602 lb. milk 1 year
Pietertje 2d	30,318 lb. milk 1 year
Boukje	21,679 lb. milk 1 year
Sultana	22,042 lb. milk 1 year
Princess of Wayne	29,008 lb. milk 1 year

Fat records were not made of the above.

[1] Hoard's *Dairyman*, Vol. XL, p. 696.

The highest official records for a year up to the present are as follows: —

	Age	Lb. Milk	Lb. Fat
Mature Cows			
Colantha 4th Johanna	8	27,432	998
Pontiac Artis	6	21,834	861
Belle Netherland Johanna	6	20,516	808
Missouri Chief Josephine	8	26,861	741
Pontiac Pleione	7	24,820	740
4 to 5 yrs. old			
Johanna DeKol Wit	..	16,176	626
Inka Plum DeKol	..	19,398	626
3 to 4 yrs. old			
Pietertje Lass 2d's Johanna	..	18,134	633
Katy Gerben	..	18,573	620
2 to 3 yrs. old			
Copia Heng. 2d's Buttercup	..	18,349	679
K. P. Alcarta	..	15,528	609

SCALE OF POINTS

FOR HOLSTEIN-FRIESIAN BULL

Head — Showing full vigor; elegant in contour	2
Forehead — Broad between the eyes; dishing	2
Face — Of medium length; clean and trim, especially under the eyes; the bridge of the nose straight	2
Muzzle — Broad with strong lips	1
Ears — Of medium size; of fine texture; the hair plentiful and soft; the secretions oily and abundant	1
Eyes — Large; full; mild; bright	2
Horns — Short; of medium size at base; gradually diminishing towards tips; oval; inclining forward; moderately curved inward; of fine texture; in appearance waxy	1
Neck — Long; finely crested (if the animal is mature); fine and clean at juncture with the head; nearly free from dewlap; strongly and smoothly joined to shoulders	5

Shoulders — Of medium height; of medium thickness, and smoothly rounded at tops; broad and full at sides; smooth over front 4
Chest — Deep and low; well filled and smooth in the brisket; broad between the forearms; full in the foreflanks (or through at the heart) 7
Crops — Comparatively full; nearly level with the shoulders . 4
Chine — Strong; straight, broadly developed, with open vertebræ 6
Barrel — Long; well rounded; with large abdomen; strongly and trimly held up 7
Loin and Hips — Broad; level or nearly level between hook bones; level and strong laterally; spreading out from the chine broadly and nearly level; the hook bones fairly prominent 7
Rump — Long; broad, high; nearly level laterally; comparatively full above the thurl; carried out straight to dropping of tail 7
Thurl — High; broad 4
Quarters — Deep; broad; straight behind; wide and full at sides; open in the twist 5
Flanks — Deep; full 2
Legs — Comparatively short; clean and nearly straight; wide apart; firmly and squarely set under the body; arms wide, strong and tapering; feet of medium size, round, solid and deep 5
Tail — Large at base, the setting well back; tapering finely to switch; the end of bone reaching to hocks or below; the switch full 2
Hair and Handling — Hair healthful in appearance; fine, soft and furry; skin of medium thickness and loose; mellow under the hand; the secretions oily, abundant and of a rich brown or yellow color 10
Mammary Veins — Large; full; entering large orifices; double extension; with special development, such as forks, branches, connections, etc. 10
Rudimentary Teats — Large; well placed 2
Escutcheon — Largest; finest 2

100

FOR HOLSTEIN-FRIESIAN COW

Head — Decidedly feminine in appearance; fine in contour .	2
Forehead — Broad between the eyes; dishing	2
Face — Of medium length; clean and trim especially under the eyes, showing facial veins; the bridge of the nose straight	2
Muzzle — Broad with strong lips	1
Ears — Of medium size; of fine texture; the hair plentiful and soft; the secretion oily and abundant	1
Eyes — Large; full; mild; bright	2
Horns — Small; tapering finely towards the tips; set moderately narrow at base; oval; inclining forward; well bent inward; of fine texture; in appearance waxy	1
Neck — Long; fine and clean at juncture with the head; free from dewlap; evenly and smoothly joined to shoulders .	4
Shoulders — Slightly lower than hips; fine and even over tops; moderately broad and full at sides	3
Chest — Of moderate depth and lowness; smooth and moderately full in the brisket, full in the foreflanks (or through the heart)	6
Crops — Moderately full	2
Chine — Straight; strong; broadly developed, with open vertebræ	6
Barrel — Long; of wedge shape; well rounded; with a large abdomen, trimly held up (in judging the last item age must be considered)	7
Loin and Hips — Broad; level or nearly level between the hookbones; level and strong laterally; spreading from chine broadly and nearly level; hookbones fairly prominent	6
Rump — Long; high; broad with roomy pelvis; nearly level laterally; comparatively full above the thurl; carried out straight to dropping of tail	6
Thurl — High; broad	3
Quarters — Deep; straight behind; twist filled with development of udder; wide and moderately full at the sides . .	4
Flanks — Deep; comparatively full	2
Legs — Comparatively short; clean and nearly straight; wide apart; firmly and squarely set under the body; feet of medium size, round, solid and deep	4

Tail — Large at base, the setting well back; tapering finely to switch; the end of the bone reaching to hocks or below; the switch full 2

Hair and Handling — Hair healthful in appearance; fine, soft and furry; the skin of medium thickness and loose; mellow under the hand; the secretions oily, abundant, and of a rich brown or yellow color 8

Mammary Veins — Very large; very crooked (age must be taken into consideration in judging of size and crookedness); entering very large or numerous orifices; double extension; with special developments such as branches, connections, etc. 10

Udder and Teats — Very capacious; very flexible; quarters even; nearly filling the space in the rear below the twist, extending well forward in front; broad and well held up . 12

Teats — Well formed; wide apart, plumb and of convenient size 2

Escutcheon — Largest; finest 2

———

100

CHAPTER V

THE CHANNEL ISLAND BREEDS

JERSEYS

Origin and Distribution in Europe. — The Jersey and the Guernsey breeds are often called the Channel Island breeds. They take their name from the islands of the same names, which are a part of the group called the Channel Islands. This group of several small islands lies in the entrance to the English Channel about nine miles from the coast of France and about seventy from England.

The cattle from these islands were formerly classed together and called Alderney, after the third island in size of the group. The cattle on these islands are supposed to be descendants of the cattle of Normandy and Brittany in France. According to Kellar, they belong to the *Bos sondaicus* type, and are therefore related in origin to the Brown Swiss, the Devons, Kerry, and more or less to other cattle of England, but not to the Holsteins. When they were brought from France to the Islands is not known, but it is known that they have been kept pure for a very long time. Since 1789 a law has been in force on Jersey Island which entirely prohibits the importation of cattle except for slaughter. A few years later Guernsey adopted a similar law.

Outside of Jersey Island this breed is found quite numerous

in England, but only a few are used elsewhere in Europe. The first demand for them outside of their native island came from England where they were placed on the estates of the nobility largely on account of their beauty. Even at the present time this breed does not contribute very much to the total dairy products of England.

Conditions on the Isle of Jersey. — Jersey Island is eleven miles long and about nine wide. Its area is 36,680 acres, of which 25,000 are tillable. The population is about 60,000. This island rises from the level of the ocean on the south in a long gradual slope to the north side, which has cliffs about 200 feet high along the ocean. The climate is mild and even, grass remains green throughout the year and is rather fine and nutritious. The cattle are pastured during the day by the tethering system. From May to October the cows, as a rule, remain outdoors all the time. In winter the cows are out in the daytime and in the evening are housed and fed hay, roots and a small ration of bran, or oil cake. But little grain is fed at any time.

Two crops at least of some kind are raised each year on the same land. The average annual rental is over $50 per acre, and this includes ground occupied by the dwelling house and barn as well as for the cultivated land. The Island was at one time in a very low state of fertility, but its productiveness has been increased until it ranks among the highest developed agricultural regions in Europe. About 10,000 cows are kept on the Island, or one to every 2.2 acres of cultivated land.

The cattle have been bred and improved with special reference to butter production for about 100 years. In 1834 a scale of points was made out for both cows and bulls and prizes offered for the animals conforming nearest to the scale of points. The breed began to improve more rapidly from that time on. At the present time the cattle on the island are a very uniform lot, but their average production is probably lower than that of equally good representatives of the breed in America, largely because they are fed a less liberal ration, especially of grain.

Plan of Registration on the Island. — The plan of registration of cattle on the Island is quite different from that followed in America. Cows are registered as Pedigree Stock and as Foundation Stock; bulls as Pedigree Stock only. Within twenty-four hours after the birth of a heifer calf which is to be registered, the owner must notify a member of the Agricultural Department, who issues a certificate showing the calf is from the cow claimed. This certificate, with one from the owner of the sire, is filed with the Secretary of the Register within six months. This is called preliminary registration.

Every two months examinations are held for the qualification of these registered cattle. The cattle to be examined are taken to the appointed place and examined by the judges. If the heifers meet with the approval of the judges, they are given qualification C., if commended, or if of exceptional merit, they are given qualification H. C., high commended. Cows in milk not registered under the first qualification may be examined for foundation stock, and if passed, are registered with the others. When a bull comes up for examination, his dam must be shown also, and her qualifications are taken into account before registering the bull. Animals passing these examinations are given herd book numbers.

Practically no milk or butter records are kept except one-day tests at fairs. The system of registration followed has resulted in making the Island animals a strikingly uniform and beautiful lot, but does not tend to the most rapid development of the milking qualities. The testing system in use by the American Jersey Cattle Club for testing the milk and butter producing capacity of cows is in advance of the system practiced on Jersey Island as far as improving the dairy qualities of the breed are concerned, while the Jersey Island plan is certain to result in a more uniform type and greater beauty.

Importations to America. — In 1850 several Jersey cows were imported to Hartford, Conn., and in 1868 S. S. Stephens of Montreal, Canada, imported nine animals, from which have descended some of the most famous producers in the Jersey breed. Since 1868 the importations were numerous until about 1890, and then few were brought over for several years. At present several importations are generally made each year. The interests of the breed in the United States are looked after by the American Jersey Cattle Club. Something over 300,000 animals have been registered in the United States up to 1910, of which about one fourth are bulls.

They are found in all parts of America, but most numerously in the Eastern and Middle states. In the South they include practically all of the dairy cattle. If numbers be taken as the basis of judgment, the Jersey is the most popular breed of dairy cows in America. The popularity of this breed has been helped by the fact that it has been in this country longer than other breeds, and further by the skill shown by the American Jersey Cattle Club in looking after its interests.

At an early date this breed was afflicted by a color craze, which injured it somewhat for several years. The fad was for solid colors, which means no white markings, and for a black tongue and switch. At present very little attention is paid to color, although the majority of the Jerseys found in this country have the solid colors and black points. In the late seventies and early eighties a great boom struck the breed in the United States. Cows of the St. Lambert breeding brought enormous prices. As high as $25,000 was paid for a single cow. In 1893 twenty-five animals each of the Jersey, Guernsey, and Shorthorn breeds competed in a dairy test at the World's Fair, Chicago. In both the production of cheese and of butter the Jerseys won first place on total production and economy of production. This gave the breed greatly increased popularity, and their numbers increased very rapidly in the following years. They also stood first in the dairy test at the World's Fair in St. Louis, in 1904, with the highest average production and greatest economy of production of butter fat.

Form and Characteristics. — The Jersey is the smallest of the dairy breeds, with the exception of the Kerry. The weight of the average cow is generally between 800 and 900 pounds. The bulls, as a rule, range from 1200 to 1700 pounds. The breeders in America have generally favored the larger animals, and for this reason, and possibly also on account of the more liberal feeding practiced, it is generally believed the breed tends to gradually increase in size after a few generations in America. Cows weighing 1000 pounds are quite common here, but unknown on the Jersey Island. Recently, on account of the numerous importations and the wide use of bulls of a smaller type from the Island, the tendency for the average size of the breed to increase is checked, temporarily at least.

The difference is so marked between the imported or their near descendants, and those descended from the early importations, that two types are generally recognized, the American and the Island types. The American type is well represented by animals of the St. Lambert breeding. This type is larger and coarser than the Island type, and less beautiful. This type is often deficient in fore udder development, is inclined to coarseness in the head and pelvic region, and is often lacking in general symmetry. Cows of the American type hold most of the best milk and butter records of the breed at present.

The origin of this type is to be credited largely to Philip Dauncey, and in part to W. G. Duncan, both of England. Mr. Dauncey first made the Jersey breed well known in England through the remarkable results of his breeding operations, which covered forty-one years' time, beginning in 1826. He developed large, rather coarse animals of great constitution and with remarkable milking qualities. Mr. Duncan began breeding Jerseys in 1849, and made use of the Dauncey blood largely, and continued the development of the same type.

In developing this type, these breeders selected animals of the type they desired, then inbred freely and continuously. They also bred the heifers to calve at three years of age, which probably was an important factor in developing a large animal. The early importations to America were largely cattle from the Duncan herd, or were descendants of the Dauncey herd. The fad for solid colors, formerly so strong in America, was introduced by these breeders, and the practice of inbreeding followed so persistently by them has been continued by American breeders.

The Island type is small and delicate-looking, beautiful in form and with splendidly developed udders, especially in front. They have fine, symmetrical heads and necks, and level flat rumps. This type has been the favorite in recent years in show ring competition, and includes the most fashionable breeding and highest priced animals of the breed at present.

A good Jersey cow is the model of what is generally taught to be the dairy form. She has the pronounced wedge shape, an immense barrel for her size, a well-developed udder, and does not carry a pound of surplus fat while in full flow of milk. The color of the Jersey varies greatly. It may be any shade of yellow, except orange, from almost white to very dark squirrel gray or black. The most common color is fawn with black shadings below and on the head. White spots may appear most commonly on the underline, but they are not generally looked upon with favor, especially among breeders of the American type. The tongue and switch are generally black. The muzzle is intensely black, encircled by a light-colored ring. The bulls are as a rule darker in color than the cows.

Cows of this breed are quite sensitive, on account of having a highly developed nervous temperament, which is not very well under control. When carefully handled, they become exceedingly gentle. On the other hand, when carelessly handled or abused, they become very much the reverse. They are more easily disturbed by unusual surroundings than other dairy breeds.

Jerseys are easy keepers, but, like all dairy animals, need

good pastures and liberal feeding for good returns. They do better than the Holstein on rough and scanty pastures, but are not equal to the Ayrshire in this respect. They are best adapted for a mild climate, and better adapted than other breeds for warm southern climates.

As meat producers they stand low, even for dairy breeds. The body fat is very yellow and is not well distributed. The calves are small, weighing usually between 50 and 60 pounds at birth, and do not gain rapidly for the first few weeks. For this reason they are not well adapted for veal production. If Jersey calves are raised for meat, they should usually be sold by the time they are eight or ten months old.

This breed is very prepotent when crossed upon common cattle or grades of other breeds. The cross usually partakes strongly of the Jersey type and milking qualities. The heifers mature young. They are usually sufficiently well developed to come into milk at the age of 22 or 24 months.

Dairy Characteristics. — Reports of American experiment stations up-to-date show the following facts regarding the yield and composition of milk produced by pure-bred Jerseys in the herds belonging to these institutions: —

	Average	No. of Cows Represented
Pounds milk per year . .	5508	153
Per cent fat	5.14	154
Yield fat per year	283	153
Per cent total solids . . .	14.9	29

On the average the fat constituted 34.5 per cent of the total solids as compared with 28 per cent for the Holsteins.

In quantity of milk produced the Jersey, as a breed, is exceeded by the Holstein, Ayrshire, and probably by the Brown Swiss and Guernsey among the common dairy breeds, and in England by the milking Shorthorn. On account of this fact, where milk is sold by quantity alone, as is too often the case with market milk, at milk condenseries and at many cheese factories, the Jersey does not find her best adaptation.

As an economical producer of butter fat the Jersey and her near relative, the Guernsey, are unsurpassed. The Jersey milk has the highest per cent of fat of any dairy breed common in this country. In economy of production of fat this breed has always led where opportunity has been given to make fair comparison with other breeds. The probable explanation of this is the fact that the fat constitutes a greater proportion of the total solids than it does in other breeds. The most prominent and best known characteristics of Jersey milk are the high per cent of fat, the pronounced yellow color, and the easy creaming of the milk. The latter characteristic is due to the unusually large fat globules. The large fat globules also cause the fat to churn easily, which is something of an advantage in butter-making. The same ease of churning is a slight disadvantage when the milk is to be handled much, as in the market milk business, because it results in small masses of butter appearing on the surface.

In persistency of milking the Jersey ranks very high, probably the highest of any breed. Cows of this breed are general favorites as family cows on account of the richness of their milk, its easy creaming characteristics, their persistency of milking, their easy keeping qualities and gentleness. The Jersey cow finds her special adaptation as a family cow or as an economical producer of butter fat. In the judgment of the author a lack of vitality in the young calves and of good breeding qualities with the cows are the weakest points of this breed.

Testing System. — In 1884 the Jersey Cattle Club authorized seven-day butter tests. The first volume of these tests was published in 1889. These tests cover seven days' time and are made by the owner, who afterwards takes oath to the correctness of the results reported. Cows producing 14 pounds or over of butter in seven days are admitted to this registry, and such cows are afterwards spoken of as tested cows, or said to be in the "14 pound list." These early private butter records, while probably in the main correct, are not looked upon by the public with the same confidence as are those made more recently under official supervision. The highest among the private seven-day records is Princess 2d, who is credited with a production of 46 pounds of butter in seven days from 299 pounds of milk. If this was normal butter containing at least 82.5 per cent of fat, the milk must have contained 13.7 per cent fat. Since the Babcock test has been in use no cow has been found in this breed testing much over 7 per cent when any quantity of milk is produced.

In 1903 the Register of Merit was established and the rules were changed to admit records made by using the Babcock test.

Bulls are entered in two classes. To be eligible to Class A a bull must have a score of 80 points and have three daughters from as many different dams in the Register of Merit. Bulls are eligible to Class B without scoring when they have three daughters entered. There are four classes of cows. To be entered in Class AA a cow must: —

(*a*) Meet the requirements of milk and fat production.

(b) Have a score of at least 80 points.

(c) Give birth to a living class within 125 days after the close of the year's record.

Cows that have met the requirements under (a) and (b) above are entered in Class A. Class B includes those meeting the requirements under (a) alone.

In 1911 the rules were amended so that now only authenticated butter fat and milk records are accepted covering seven days or a full year.

The representative of an experiment station or other agent of the Jersey Cattle Club tests and weighs each milking separately for seven days. The cow must produce 12 pounds of fat in 7 days. This test may also cover a year, and in this case the owner weighs the milk from each milking during the year, and for two days in each month the milk is weighed and tested by the representative of an experiment station or by another agent of the Jersey Club. The average of the per cent of fat for these two days is taken as the average for the month. A two-year-old heifer must produce 260 pounds of fat per year and a mature cow 400 pounds of fat.

Prominent Jersey Families. — There are some fairly well marked families in this breed. The best known, probably, is the St. Lambert. This family originated in Canada, and is descended from the cattle imported by Stephens of Montreal and St. Clair of Vermont. The bulls Stoke Pogis and Stoke Pogis 3d are supposed to be predominant factors in the formation of this family, which includes many of the best known animals of this breed in America. Animals of this family, as a rule, are large in size and often rather coarse in make-up, generally fawn or gray in color, and seldom black. It has produced many remarkable dairy cows. Like other prominent Jersey families, it is deeply inbred.

The Island cattle imported to America within recent years mostly carry the blood of Golden Lad. Recently the descendants of Golden Fern's Lad, a grandson of Golden Lad, have been especially sought after. The bull, Golden Lad, is counted the greatest bull the Island has yet produced, and his descendants show remarkable uniformity and high quality.

Inbreeding. — The leading breeders in America have usually followed Dauncey and Duncan in practicing inbreeding freely. As a rule, every high-class animal of the American type is more or less inbred. The Island breeders do not follow this practice to such a marked extent. The object of inbreeding has been to increase the proportion of blood of certain animals supposed to be unusually prepotent in transmitting milking qualities and other desirable characteristics.

The following are the ten highest yearly records up to the present for the Jersey breed: —

	Lb. Milk	Lb. Fat
Jacoba Irene	17,253	952
Sophie 19th of Hood Farm	14,373	855
Adelaide of Beechlands	15,572	849
Rosaire's Olga 4th Pride	14,104	836
Financial Countess	13,248	795
Molle of Edgewood	14,036	705
Lass 30th of Hood Farm	11,990	685
Bessie Bates	13,895	680
Olive Dunn	9,930	671
Peers Surprise	14,452	653

SCALE OF POINTS FOR JERSEY COW

Head, 7.

  A — Medium size, lean; face dished; broad between eyes and narrow between horns 4

  B — Eyes full and placid; horns small to medium, incurving; muzzle broad, with muscular lips; strong under jaw 3

Neck, 5.

  Thin, rather long, with clean throat; thin at withers . . . 5

Body, 33.

  A — Lung capacity, as indicated by depth and breadth through body, just back of fore legs 5

  B — Wedge shape, with deep, large paunch; legs proportionate to size and of fine quality 10

  C — Back straight to hip bones 2

  D — Rump long to tail setting and level from hip bones to rump bones 8

  E — Hip bones high and wide apart; loins broad, strong . 5

  F — Thighs flat and well cut out 3

Tail, 2.

  Thin, long, with good switch, not coarse at setting on . . 2

Udder, 28.

  A — Large size and not fleshy 6

  B — Broad, level or spherical, not deeply cut between teats . 4

  C — Fore udder full and well rounded, running well forward of front teats 10

  D — Rear udder well rounded, and well out and up behind . 8

Teats, 8.

  Of good and uniform length and size, regularly and squarely placed 8

Milk Veins, 4.

  Large, tortuous and elastic 4

Size, 3.

  Mature cows, 800 to 1000 pounds 3

General Appearance, 10.

  A symmetrical balancing of all the parts, and a proportion of parts to each other, depending on size of animal, with the general appearance of a high-class animal, with capacity for food and productiveness at pail 10

                                                                                           100

SCALE OF POINTS FOR JERSEY BULL

Head, 10.

   A — Broad, medium length; face dished; narrow between horns; horns medium in size and incurving . . .   5

   B — Muzzle broad, nostrils open, eyes full and bold; entire expression one of vigor, resolution, and masculinity   5

Neck, 10.

   Medium length, with full crest at maturity; clean at throat.   10

Body, 54.

   A — Lung capacity as indicated by depth and breadth through body just back of fore shoulders; shoulders full and strong   15

   B — Barrel long, of good depth and breadth, with strong, well-sprung ribs   15

   C — Back straight to hip bones   2

   D — Rump of good length and proportion to size of body, and level from hip bone to rump bone   7

   E — Loins broad and strong; hips rounded, and of medium width compared with female   7

   F — Thighs rather flat, well cut up behind, high arched flank   3

   G — Legs proportionate to size and of fine quality, well apart, and not to weave or cross in walking   5

Rudimentary Teats, 2.

   Well placed   2

Tail, 4.

   Thin, long, with good switch, not coarse at setting on . .   4

Size, 5.

   Mature bulls, 1200 to 1500 pounds   5

General Appearance, 15.

   Thoroughly masculine in character, with a harmonious blending of the parts to each other; thoroughly robust, and such an animal as in a herd of wild cattle would likely become master of the herd by the law of natural selection and survival of the fittest   15

                                                          100

GUERNSEYS

Origin and Distribution in Europe. — The Guernsey breed originated on the island of the same name, which is the second in size of the Channel Islands. Like the Jerseys, they are presumably descended from the cattle of France. A century ago the cattle on the islands of Jersey and Guernsey were practically the same in form and color, but even then the Guernsey is said to have been a little larger. In 1819 laws were passed prohibiting the importation of cattle into Guernsey, and since that time the breed has been kept pure.

Professor Low, writing in 1841, says the cattle of Guernsey and Jersey were at that time essentially the same, although he further describes the former as being larger, the form rounder, and the bones less prominent than with the cattle on Jersey Island. He also refers to the unusually orange yellow skin and yellow milk and butter. It appears from his writings that these two breeds at that date were nearer together in type than is now the case, but that the Guernseys showed the same characteristics in general as at present.

The Guernseys have been taken to England in considerable numbers, and are used considerably, especially in the southwestern part, but on the whole are of little importance in that country from a dairy standpoint on account of their comparatively small numbers. They are not found elsewhere in Europe to any extent.

Conditions in Guernsey. — The conditions on Guernsey Island are practically the same as in Jersey. The Island is second in size of the Channel group, and has a population of about 41,000. It is ten miles long, with an area of about 16,000 acres, of which 12,000 are tillable. The climate is a little more severe than that of Jersey, as it is exposed toward the northwest. The south coast rises from 200 to 400 feet above the ocean, and slopes to the level of the ocean on the north.

The system of agriculture and the general management of the cattle is much the same as that followed in Jersey. There are about 8000 cattle on the Island. The Guernsey breeders have kept the idea of utility alone in view, and have given less attention to the development of the beauty of their breed.

Importation to America. — There seem to be no records available in regard to the first animals of this breed brought to America. Most of the importations were made in the period from 1880 to 1890, or since 1900. The American Guernsey Club was founded in 1877. Up to the present there have been something like 50,000 animals registered in the Guernsey Register, the first volume of which was issued in 1878. Of these close to one third are bulls.

Guernseys are distributed most abundantly in the East, especially in New York, Massachusetts, and Pennsylvania, and in the West in Wisconsin and Minnesota. Their rapid increase in popularity within recent years has been brought about to no small extent by the excellent system of yearly records established by the Guernsey Cattle Club and by the remarkable records made. The breeders of this breed should also be commended for the general use they have made of the tuberculin test, probably standing first in this respect in America.

Form and Characteristics. — The Guernsey cow weighs about 1000 pounds on the average, ranking in size about the same as the Ayrshire, and at least 100 pounds larger than the Jerseys. This breed is coarser boned and more irregular in conformation than the Jersey. The colors resemble the Jersey in general, but include some colors not found in that breed. The common colors are a reddish yellow, or lemon or orange fawn, with white markings. The white markings are usually found on the face, flanks, legs, and switch, but may be on any part of the body.

The temperament of the Guernsey cow is good. She is alert and wide awake, but not nervous and irritable. She has a good dairy conformation, and gives the impression of a plain animal bred for utility. While the breed lacks uniformity in type to some extent, there are no special types of the breed recognized. The breeding qualities are fair, but probably not equal to some other breeds. As a breed they do not mature quite so early as does the Jersey. The heifers come into milk at 22 to 26 months. Like the Jersey, they have little adaptation for meat production.

Dairy Characteristics. — In yield of milk the Guernsey ranks with the Jersey. The experiment stations report the following results, which are all yearly records of animals owned by these institutions: —

Average		No. of Cows Represented
Milk per year	5509	17
Per cent fat	4.98	21
Yield fat per year	274	17
Per cent total solids	14.2	6

In the records above reported the fat constituted 35 per cent of the solids as compared with 34.5 per cent for the Jersey and 28 per cent for the Holsteins.

Guernsey milk and butter has the yellowest color of any breed. For this reason milk from this breed is often mixed with that from other breeds for the sake of the high color it imparts. The color of Guernsey butter is so high that it is even occasionally objected to in the market by those not familiar with it, especially when the cows are on fresh pasture. The fat globules are likewise the largest on the average of any breed. The Guernseys have the same advantages and disadvantages as the Jerseys as cows for producing market milk, and on the whole they cannot be said to be well adapted for this purpose, although special favorites for producing a high-grade cream for market. They share with the Jerseys the power of producing butter fat very economically on account of the richness of their milk. For this reason they find their greatest adaptation as producers of butter or cream.

An excellent system of testing for advanced registration was adopted by the Guernsey Cattle Club in 1901. The test at first was for either seven days or a year, but now only the latter is accepted. It is under the supervision of an experiment station or agricultural college. The records include milk produced and butter fat as determined by the Babcock method. The owner of the cow keeps the milk records in detail throughout the year, and the supervisor of the test visits the herd once each month and weighs and tests each milking for two days. The average per cent of fat for these two days is taken as the average for the entire month. The cows are admitted to Advanced Registry by meeting the following requirements: —

A cow beginning at two years of age must produce 6000 pounds of milk within a year, with an addition of 3.65 pounds of milk per day up to five years of age, at which time the requirements are 10,000 pounds of milk. The butter fat requirements for admission are the production of 250.5 pounds fat, if the record begins at two years of age, and for each day past two years at the time of beginning .1 pound is added. The requirement for cows of mature age is 360 pounds of fat per year.

Bulls are admitted to Advanced Registry by having two daughters in the Registry.

The Guernseys have a remarkable list of yearly butter fat records. The following are the highest records in print up to the present: —

	Lb. Milk	Lb. Fat
Mature Cows		
Yeksa Sunbeam	14,920	857
Dolly Bloom	17,297	836
Imp. Princess Rhea	14,009	775
Modena	14,001	728
4 to 5 Years Old		
Imp. Itchen Daisy	13,636	714
Standard Morning Glory	12,917	714
3 to 4 Years Old		
Dolly Dimple	18,458	906
Dairy Maid of Pinehurst	14,562	860
2 to 3 Years Old		
Dolly Dimple	14,562	703
Langwater Princess	12,280	651

SCALE OF POINTS FOR GUERNSEY COWS

Dairy Temperament. Constitution, 38.	Clean-cut, lean face; strong, sinewy jaw; wide muzzle with wide-open nostrils; full, bright eye with quiet and gentle expression; forehead long and broad	5
	Long, thin neck with strong juncture to head; clean throat. Backbone rising well between shoulder blades; large, rugged spinal processes, indicating good development of the spinal cord	5
	Pelvis arching and wide; rump long; wide, strong structure of spine at setting on of tail. Long, thin tail with good switch. Thin, incurving thighs	5
	Ribs amply and fully sprung and wide apart, giving an open, relaxed conformation; thin arching flanks	5
	Abdomen large and deep, with strong muscular and navel development, indicative of capacity and vitality	15
	Hide firm yet loose, with an oily feeling and texture, but not thick	3
Milking Marks, denoting Quantity of Flow, 10.	Escutcheon wide on thighs; high and broad, with thigh oval	2
	Milk veins long, crooked, branching, and prominent, with large or deep wells	8
Udder Formation, 26	Udder full in front	8
	Udder full and well up behind	8
	Udder of large size and capacity	4
	Teats well apart, squarely placed, and of good and even size	6
Indicating Color of Milk, 15	Skin deep yellow in ear, on end of bone of tail, at base of horns, on udder, teats, and body generally. Hoof, amber-colored	15
Milking Marks denoting Quality of Flow, 6	Udder showing plenty of substance, but not too meaty	6
Symmetry and Size, 5	Color of Hair, a shade of fawn, with white markings. Cream-colored nose. Horns amber-colored, small, curved and not coarse	3
	Size for the breed: Mature cows, four years old or over, about 1050 lb.	2
		100

SCALE OF POINTS FOR GUERNSEY BULLS

Dairy Temperament. Constitution, 38	Clean-cut, lean face; strong sinewy jaw; wide muzzle with wide-open nostrils; full, bright eye with quiet and gentle expression; forehead long and broad	5
	Long, masculine neck with strong juncture to head; clean throat. Backbone rising well between shoulder blades; large, rugged spinal processes, indicating good development of the spinal cord	5
	Pelvis arching and wide; rump long; wide, strong structure of spine at setting of tail. Long, thin tail with good switch. Thin, incurving thighs	5
	Ribs amply and fully sprung and wide apart, giving an open relaxed conformation; thin arching flank	5
	Abdomen large and deep, with strong muscular and navel development, indicative of capacity and vitality	15
	Hide firm yet loose, with an oily feeling and texture, but not thick	3
Dairy Prepotency, 15	As shown by having a great deal of vigor, style, alertness, and resolute appearance	15
Rudimentaries and Milk Veins, 10	Rudimentaries of good size, squarely and broadly placed in front of and free from scrotum. Milk veins prominent	10
Indicating Color of Milk in Offspring, 15	Skin deep yellow in ear, on end of bone of tail, at base of horns and body generally, hoofs amber-colored	15
Symmetry and Size, 22	Color of hair, a shade of fawn with white markings. Cream-colored nose. Horns amber-colored, curving and not coarse	8
	Size for the breed: Mature bulls, four years old or over, about 1500 lb.	4
	General appearance as indicative of the power to beget animals of strong dairy qualities	10
		100

CHAPTER VI

AYRSHIRES

ORIGIN AND CHARACTERISTICS

Origin and Distribution in Europe. — The native home of this breed is the county or shire of Ayr in southwest Scotland. It is comparatively a new breed, but has made wonderful advancement in a short time. The origin of this breed is veiled in some obscurity, but traces back to the latter part of the eighteenth century. The cattle of this district are described in 1750 as being small, ill-fed, ill-shaped, and producing but little milk. In color they were black and white. This foundation stock was probably descended from or kin to the original wild white cattle described in early historic records and represented now by the wild park cattle.

During the latter half of the eighteenth century a widespread movement for better cattle spread over Great Britain, and resulted in an immense improvement being made in the cattle of Ayrshire as elsewhere. This improvement was brought about by more careful selection and breeding, and especially by the introduction of blood from several other regions. According to Howard,[1] writing in 1863, cattle imported from Holland furnished one of the most important crosses made. Another important cross was made with cattle

[1] Vol. 1, Herd Record of the Association of Breeders of Thoroughbred Neat Cattle.

of the Durham or Teeswater breed, which afterwards in the improved form became known as the Shorthorns. It is reasonably certain, according to this author, that Channel Island blood was also introduced quite generously. It is also probable that Holderness blood was introduced to some extent at an early date.

Professor Low, the eminent English authority, makes the following statement: " From all evidence which, in the absence of authentic documents, the case admits of, the dairy breed of Ayrshire cows owes its character to a mixture of the blood of the races of the continent and of the dairy breed of Alderney."

In Ayrshire some attention has always been given to the beef-making capacity of the breed especially in the early period of their development. During the early part of the nineteenth century, the hind quarters especially were improved and the udder brought to its present symmetrical proportions.

They are the leading dairy cattle in Scotland, and are common in some parts of England. This is one of the few British breeds that has spread to any extent on the Continent of Europe. They are quite numerous in Finland, Sweden, and Norway, while in New Zealand they are the most important dairy breed.

Conditions in Ayrshire. — Ayrshire is situated on the southwest coast of Scotland. The land rises from the level of the ocean on the west to mountains about 2000 feet high on the east. The soil is a heavy clay of moderate fertility spread over a hilly surface. The temperature ranges from 25 to 65° F. during the year, and is not subject to great extremes, although swept by fierce storms occasionally. The moist air and abundant rainfall results in good pastures. The milk is used mostly for cheese making.

Importation and Distribution in America. — Ayrshires were brought to Canada in the early part of the nineteenth century by Scotch settlers. More recently the importation of breeding stock into Canada has been extensive. The first importation into the United States was probably in 1837, when several were brought to Massachusetts by the Society for the Promotion of Agriculture, although it is claimed that some were taken to Connecticut as early as 1822. These importations continued at intervals for twenty or thirty years, then gradually ceased on account of the serious objection raised to the short teats of the imported animals. The Ayrshires now known as the American type are descended from these early importations and their improved descendants.

Within recent years a considerable number of cattle of this breed have been imported into the United States. The importations into Canada have also been numerous in recent years, so that the Canadian animals are largely of the type now called the Scotch type. This type is now the most popular among the American breeders also.

The Ayrshires have never been boomed or even well advertised in the United States, and their increase in popularity has therefore been entirely on their merits. This breed took part in the dairy demonstration at the Pan-American Exposition in Buffalo in 1898, and made a very creditable record, standing third as a breed in butter produced and in net profit. The first herd book of the breed was established in the United States in 1863, the Scotch Herd Book began in 1878, and the Canadian in 1870.

Form and Characteristics. — In size, the Ayrshires rank between the Jersey and Holstein breeds. The average cow weighs about 1000 pounds at maturity, while some exceed this figure considerably. The bulls range from 1400 to 2000 pounds. The Scotch type is probably rather under these figures.

The common color is spotted red or brown and white in varying proportions. In the American type the red or brownish red usually predominates, with only a small amount of white, while in the Scotch type the white predominates. The two colors are distinct, and do not blend to form a roan. The horns are rather long, and as a rule curve outwards and upwards, and in some cases slightly backwards. The Scotch type have rather heavy horns, especially with the bulls.

In disposition they stand rather between the Jersey and Holstein. They are more active and alert than the Holstein, and like them are less affected by unusual surroundings than some other breeds. In form the Ayrshire do not show the extreme angular dairy type as exhibited by high-class Jerseys or Holsteins. They are smoother over the shoulders, back, and hips, and have fuller rear quarters. At the same time, the barrel is large, showing great capacity, and the udder development the most perfect of any breed. For many years the Scotch breeders have bred especially for large, symmetrical udders, and have attained this end with remarkable success. The udder is attached high behind, and extends far forward, with a flat, even lower surface. The teats are placed regularly on the udder, and are of uniform size. Ayrshires are regular and sure breeders. This probably results from the favorable conditions under which the breed has been developed and from the avoidance of inbreeding.

There are two fairly well marked types, but with individuals standing all the way between the two types. These two types are known as the American and Scotch. As already explained, the former are descended from the early importations dating before 1850. The cattle of these early importations were objected to on account of their short teats, and the breeders set about to remedy this defect by selection and breeding. This they accomplished, and at the same time increased the size somewhat and retained the other desirable characteristics. In color this type is mostly red or brownish red, with a few white spots. They do not show, as a rule, as symmetrical an udder as the imported animal, and are more of the angular, spare, dairy form, and have longer teats.

The Scotch type is that bred at present in Ayrshire and imported to this country in large numbers within recent years. In color these animals are mostly white with brown ears, and a few red or brown spots irregularly arranged on the body. They are more handsome and symmetrical than the American type animals, and have a remarkable udder development, as already described. This type is best represented in Canada at present, but is being rapidly introduced into the United States. The chief objection urged against this type is the short teats, which are often altogether too small for men to milk with convenience. In Scotland the milking is largely done by women, and the short teats are not considered objectionable. American breeders are now giving attention to increasing the length of the teats to remove this criticism.

The Ayrshire is not as early maturing as the Jersey, ranking about with the Holstein in this respect. The cows come into milk from 24 to 30 months of age. They are noted for being productive to a greater age than most breeds.

For veal production the Ayrshire ranks between the Jersey and the Holstein. The calves weigh 60 to 75 pounds at birth, and are strong and vigorous. In beef production this breed ranks high for a dairy breed. When dry they fatten readily, and are said to make a good quality of beef.

Dairy Characteristics. — The following summary of yearly records of pure-bred Ayrshire cattle owned by American experiment stations shows the chief facts regarding their milking qualities: —

Average		No. Animals Represented
Pounds milk per year	6533	24
Per cent fat	3.85	24
Yield of fat per year	252	24
Per cent total solids	12.98	17

In the above the fat constitutes 29.6 per cent of the total solids, as compared with 34.5 for the Jerseys and 28 per cent for the Holsteins.

As a breed the Ayrshires are noted for a good uniform production of milk rather than for remarkable records. They have not yet made the enormous records of fat and milk production that have been made by the Holsteins, Jerseys, and Guernseys, but at the same time there are few really inferior animals found in this breed, and the average production as a breed is probably equal to that of any other.

The fat averages a little under 4 per cent, and the total solids 12.5 per cent. The fat globules are small, and the milk and butter does not show much yellow color, ranking in this respect with the Holsteins. This breed is especially well adapted for the production of market milk, because they produce a large yield of milk of average composition. The small fat globules are an advantage in this connection, and the total solids are also well balanced between fat and other solids. The only objection to this milk for market purposes is the lack of a pronounced yellow color.

The milk of this breed is well adapted for cheese making on account of the small fat globules and relatively high per cent of casein, and is generally used for this purpose in their native land. They are fairly economical producers of butter fat, but as a breed are probably excelled in this respect by the Jerseys and Guernseys. When their milk is creamed by gravity the loss in the skim milk is greater than from the milk of the Channel Island breeds with their large fat globules, but with the cream separator this difference is too small to be taken into account.

An Advanced Register for this breed was established in 1902 to encourage and record records of dairy performances. The rules make the following requirements for admission: —

As a two-year-old a cow must produce 6000 pounds of milk and 214 pounds of fat in one year. For each day above two years there is added to the two-year-old requirements 2.74 pounds of milk and .12 pound of fat. This is continued until five years of age is reached when the cow is supposed to be mature. The requirements for this age or older are 8500 pounds of milk and 322 pounds of fat in a year.

The milk records are kept by the owner. The milk from each cow is weighed and tested for two consecutive days each month by a representative of the experiment station in that state.

A Home Dairy Test was established in 1901. In this competition prizes are given to the best herds of five and also for best individual cows. The record of milk is kept by the owner and the samples taken and records verified in the same manner as for the Advanced Register. Bulls are eligible to entry in Advanced Registry by having four daughters in the registry from as many dams or by having two daughters and scoring 80 points.

The leading official records of this breed up to the present are as follows: —

Mature cows — Netherhall Brownie 9th, 18,110 pounds of milk and 781 pounds of fat.

Rena Ross, 15,072 pounds of milk and 644 pounds of fat.

Four-year-old — Bessie of Rosemont, 14,102 pounds of milk and 578 pounds of fat.

Three-year-old — Mattie of Sand Hill, 14,381 pounds of milk and 612 pounds of fat.

Two-year-old — Hazel of Sand Hill, 11,078 pounds of milk and 627 pounds of fat.

SCALE OF POINTS FOR AYRSHIRE COW

Head, 10.
 Forehead — Broad and clearly defined 1
 Horns — Wide set on and inclining upward 1
 Face — Of medium length, slightly dished, clean cut, showing veins 2
 Muzzle — Broad and strong without coarseness, nostrils large 1
 Jaws — Wide at the base and strong 1
 Eyes — Full and bright with placid expression 3
 Ears — Of medium size and fine, carried alert 1
Neck — Fine throughout, throat clean, neatly joined to head and shoulders, of good length, moderately thin, nearly free from loose skin, elegant in bearing 3
Fore Quarters, 10.
 Shoulders — Light, good distance through from point to point, but sharp at withers, smoothly blending into body 2
 Chest — Low, deep, and full between and back of forelegs . 6
 Brisket — Light 1
 Legs and Feet — Legs straight and short, well apart, shanks fine and smooth, joints firm; feet medium size, round, solid, and deep 1

Body, 13.
- Back — Strong and straight, chine lean, sharp, and open-jointed 4
- Loin — Broad, strong, and level 2
- Ribs — Long, broad, wide apart, and well sprung 3
- Abdomen — Capacious, deep, firmly held up with strong muscular development 3
- Flank — Thin and arching 1

Hind Quarters, 11.
- Rump — Wide, level, and long from hooks to pin bones, a reasonable pelvic arch allowed 3
- Hooks — Wide apart and not projecting above back nor unduly overlaid with fat 2
- Pin Bones — High and wide apart 1
- Thighs — Thin, long, and wide apart 2
- Tail — Long, fine, set on a level with the back 1
- Legs and Feet — Legs strong, short, straight when viewed from behind and set well apart; shanks fine and smooth, joints firm; feet medium size, round, solid, and deep ... 2

Udder — Long, wide, deep, but not pendulous, nor fleshy; firmly attached to the body, extending well up behind and far forward; quarters even; sole nearly level and not indented between teats, udder veins well developed and plainly visible 22

Teats — Evenly placed, distance apart from side to side equal to half the breadth of udder, from back to front equal to one third the length; length $2\frac{1}{4}$ to $3\frac{1}{4}$ inches, thickness in keeping with length, hanging perpendicular and not tapering 8

Mammary Veins — Large, long, tortuous, branching and entering large orifices 5

Escutcheon — Distinctly defined, spreading over thighs and extending well upward 2

Color — Red of any shade, brown, or these with white; mahogany and white, or white; each color distinctly defined. (Brindle markings allowed, but not desirable.) . 2

Covering, 6.
- Skin — Of medium thickness, mellow and elastic 3
- Hair — Soft and fine 2
- Secretions — Oily, of rich brown or yellow color 1

Style — Alert, vigorous, showing strong character; temperament inclined to nervousness, but still docile 4

Weight at maturity not less than one thousand pounds ... 4

Total 100

SCALE OF POINTS FOR AYRSHIRE BULL

Head, 16.
    Forehead — Broad and clearly defined 2
    Horn — Strong at base, set wide apart, inclining upward . . 1
    Face — Of medium length, clean cut, showing facial veins . 2
    Muzzle — Broad and strong without coarseness 1
    Nostrils — Large and open 2
    Jaws — Wide at the base and strong 1
    Eyes — Moderately large, full, and bright 3
    Ears — Of medium size and fine, carried alert 1
    Expression — Full of vigor, resolution, and masculinity . . 3

Neck, 10.
    Of medium length, somewhat arched, large, and strong in the muscles on top, inclined to flatness on sides, enlarging symmetrically towards the shoulders, throat clean and free from loose skin 10

Fore Quarters, 15.
    Shoulders — Strong, smoothly blending into body with good distance through from point to point and fine on top . 3
    Chest — Low, deep, and full between back and forelegs . . 8
    Brisket — Deep, not too prominent, and with very little dewlap 2
    Legs and Feet — Legs well apart, straight, and short, shanks fine and smooth, joints firm, feet of medium size, round, solid, and deep 2

Body, 18.
    Back — Short and straight, chine strongly developed and open-jointed 5
    Loin — Broad, strong, and level 4
    Ribs — Long, broad, strong, well sprung, and wide apart . 4
    Abdomen — Large and deep, trimly held up with muscular development 4
    Flank — Thin and arching 1

Hind Quarters, 16.
    Rump — Level, long from hooks to pin bones 5
    Hooks — Medium distance apart, proportionately narrower than in female, not rising above the level of the back . . 2
    Pin Bones — High, wide apart 2
    Thighs — Thin, long, and wide apart 4
    Tail — Fine, long, and set on a level with back 1

Legs and Feet — Legs straight, set well apart, shanks fine
and smooth; feet medium size, round, solid, and deep,
not to cross in walking 2
Scrotum — Well developed and strongly carried 3
Rudimentaries, Veins, etc. — Teats of uniform size squarely
placed, wide apart and free from scrotum; veins long,
large, tortuous, with extensions entering large orifices; escutcheon pronounced and covering a large surface . . . 4
Color — Red of any shade, brown, or these with white, mahog-
any and white, or white; each color distinctly defined . . 3
Covering, 6.
Skin — Medium thickness, mellow and elastic 3
Hair — Soft and fine 2
Secretions — Oily, of rich brown or yellow color 1
Style — Active, vigorous, showing strong masculine character,
temperament inclined to nervousness, but not irritable or
vicious 5
Weight at maturity not less than 1500 pounds 4

Total 100

CHAPTER VII

BROWN SWISS

Origin. — There are two distinct and leading breeds of cattle in Switzerland and several minor ones closely related to them. In the western portion, including the region around Berne, the Simmenthaler or Fleckvieh is the common breed. This is a large-boned, spotted red and white breed used for milk and beef as well as for work. They have not been imported into this country. The other Swiss breed, known in Switzerland as the Braunvieh or Schwyz and in America as the Brown Swiss, is found in the northeast part in the cantons of Zurich, St. Galen, Luzern, and Schwyz.

This breed is probably one of the oldest in existence and is supposed to be descended from the cattle which have been used in this locality since before historic records began. According to Keller this breed represents the *Bos sondaicus*, the smaller of the two types of wild cattle supposed to have been domesticated. Bones found in the ruins of the Swiss Lake Dwellers, which date back to the Bronze Age, show a type of cattle which are apparently closely related to the present Swiss cattle. It is not believed that any considerable infusion of foreign blood ever has been introduced. The Brown Swiss is one of the few breeds originated on the Continent of Europe that is used outside of its native locality. This breed has spread over about one half of Switzerland and largely into the noted dairy districts of Algau in Bavaria.

They are also in great demand in Germany, Hungary, and other parts of Europe. Foreign buyers take all the surplus stock offered for sale. The cows bring from $125 to $150 to be used as milkers.

Conditions in Switzerland. — The total area of Switzerland amounts to 15,976 square miles, of which about 71 per cent is said to be productive. The total number of cows is approximately three fourths of a million and from these are produced dairy products sufficient for a population of three and a half million and surplus dairy products exported to a value of around $26,000,000 per year. The Brown Swiss cattle are found from the shores of Lake Constantine at an elevation of 1400 feet above the sea to near the line of perpetual snow in the Alps. During the winter season the cattle are kept in the valleys and stabled closely in rather warm but dark, poorly ventilated barns and in not especially sanitary conditions. During this season they are fed almost entirely upon hay grown in the meadows in the valleys which is handled with the greatest care and is of exceptional quality. In addition they are fed turnips, potatoes, and a small quantity of grain, usually oil cake, but never more than three or four pounds per day. As spring opens they are taken to the edge of the valleys and the lower Alps up to an elevation of about 3000 feet where they are pastured on an average of about 116 days.[1] As the summer advances the herds are moved upwards to the Middle Alps which are at an elevation of from 3000 to 6500 feet, where they are pastured on an average of 92 days. In July and August they are taken on to higher pastures which are called the High Alps at an elevation of from 6000 to 8500 feet. As winter

[1] Baechler-Massuger, *Journal British Dairy Farmers' Assoc.*, Vol. 10, p. 47.

approaches the cattle are taken to the lower levels. During the summer season they are out of doors practically all the time, being sheltered at night during the cold weather and in case of storms in the temporary shelter sheds which are built on the mountains. The herds are accompanied by the herders who remain with them and who take the milk to the cheese makers. During the summer the milk is used almost exclusively for cheese making, but during the winter is used in part for butter making. The calves are usually born in the spring in the herds that pasture in the mountains, but in the herds that remain in the valleys, the cows often freshen in the fall.

Importations to America. — The first cattle of this breed to be imported to the United States were brought by H. M. Clark of Belmont, Mass., in 1869. In 1882 an importation was made by Scott & Harris of Massachusetts. A number of importations have been made since this time, but never in very large numbers. Among the recent important importations was one made in 1904 by McLaury Brothers, of New York. At the present time they are found in almost every state, but not in large numbers. The best known herds at the present time are in Connecticut, New York, and Illinois. Their spread has not been very rapid, partly because they have not been extensively advertised. Their increase in numbers and popularity has been entirely on their own merits.

Form and Characteristics. — In appearance, animals of this breed are plain, substantial, and well proportioned, although rather fleshy, and give the impression of being somewhat coarse in the bone and in general make-up. The head and neck especially are large, contrasted with the English breeds of cattle, and seem rather plain and coarse. The back is well developed and the hair abundant and soft. As a rule, the skin is of unusually fine quality. The hind quarters are full, round, and inclined to be distinctly beefy. The cows have large, well-shaped udders with teats of sufficient size to be milked conveniently. Milk veins and milk wells are of medium development. In size the cows reach an average weight of about 1200 pounds and the bulls from 1600 to 2000 pounds. The color varies considerably in shade. They are called brown, but the prevalent color is more of a mouse color. The brown varies from a silver gray or light brown to a dark brown or almost black. The nose, switch, tongue, and horn tips are always black. The mouth is surrounded by a mealy ring, and a light stripe is always found along the backbone. Small patches of white on the under line near the udder are not objectionable, but white on other parts of the body is not desirable and in Switzerland disqualifies the animal. The uniformity in color markings adds to the attractiveness. The head, neck, and legs may be almost black.

In disposition this breed is especially good, being quiet and docile and easily handled. There are no distinct types of this breed recognized as is the case with some other breeds, but there is considerable variation in types, depending upon where they were raised and for what purpose. Those raised in the higher altitudes are said to be somewhat smaller than those raised in the valleys. Some herds have been kept largely for the purpose of raising work animals, and this has resulted in these herds being of a larger and coarser type.

This breed is not very early maturing, ranking about with the Holstein breed or perhaps a little behind in this respect. In Switzerland the cows usually come into milk at about three years of age. While they are rather slow to mature, they are noted for continuing to be sure breeders until they reach an advanced age. Their excellent breeding characteristics which are one of the strong features of this breed come from the favorable conditions under which they have been kept and the sensible management.

In Switzerland this breed is considered to be a dual-purpose breed and is usually classed that way in America. However, recently the Brown Swiss breeders decided that their breed should be classed in America as a dairy breed and developed for this purpose. The animals of this breed produce a fair quality of beef, grow rapidly, and reach a good size at an early age, but they are not received very well on the market on account of the large bones and probably in part from prejudice on account of their similarity in color to the Jerseys. For production of veal they rank high as the calves are large at birth and grow rapidly.

Dairy Characteristics. — There are comparatively few figures available showing production by animals of this breed in America. One of the leading herds of this breed owned by E. M. Barton, Hinsdale, Ill., includes cows with the following 12 months' records: —

Milk	No. Cows
12,000–13,000	2
11,000–12,000	7
10,000–11,000	8
9,000–10,000	4
8,000– 9,000	10
7,000– 8,000	2

The highest fat yield in this herd is 513 pounds, while 23 are above 400 pounds. These records indicate that this breed has high dairy qualities. The majority of the cows of this breed can be counted on to produce from 5000 to 7000 pounds of milk per year with an average fat content of about 4 per cent.

There are many records in print from cattle of this breed in Switzerland. The following figures show the average milk yield for five years of fifty cows owned by an Agricultural School, Plantahof, Graubuenden, Switzerland:—

	LB.
1892	5782
1893	5500
1894	6117
1895	6307
1896	6252

The average per cent of fat during this time was 3.7 per cent. It is reported by Mr. F. H. Mason, U. S. Consul at Zurich, that 6000 cows supply the Anglo-Swiss Condensed Milk Company an average of 5115 pounds milk per year each, and that the milk received averages 3.68 per cent fat. The average per cent of dry matter is given as 12.46 in the milk received at this factory. In composition the milk of this breed is well adapted for market milk purposes containing about the proper amount of fat and solids. In natural color the milk is about the same as that of the Shorthorn breed. The milk is well adapted for cheese making, as well as the average for butter making and for feeding to calves.

No system of Advanced Registration has been adopted in this country yet. In Switzerland the interests of the breed are looked after very carefully by a coöperative association which is subsidized by the government. The animals used for breeding purposes are approved by representatives of this association which also holds fairs at intervals which serve the purpose of markets for disposing of surplus animals. The breeding herds are so well controlled by this Association and their breeding interests promoted so skillfully that the demand for their cattle has been stimulated in other parts of Europe and the surplus stock is readily disposed of at good prices. This organization gives the records of production of twenty-eight cows, the dams of bulls offered for sale, as 9179 pounds of milk with an average fat content of 3.91 per cent and of total solids 13.28 per cent.[1]

[1] Jahrbuch der Schweiz, 1904.

CHAPTER VIII

MINOR DAIRY BREEDS — DUTCH BELTED, POLLED JERSEY, KERRY, FRENCH-CANADIAN

DUTCH BELTED

This oddly colored breed originated in North Holland and has been bred for over 200 years to produce a perfect belt of pure white in the center of a coal black body. In Holland, they are called "Lahenfield" or "Lahenvelder," which means a field of white, but conveys the idea of a white body with black ends. Very little is known of their history; they are probably descended from the cattle of Holland, the ancestors of the breed known as Holstein in America.

This breed first attracted attention about 1750, but the selection and breeding for the white belt probably began long before this, possibly as far back as the sixteenth century. According to the early records, these cattle were bred by the Nobility of North Holland, on account of the peculiar markings. They also bred hogs and poultry with a white body line somewhat similar. The Lahenvelder poultry of England and America, the Lakenscheswine of Holland and Germany, and the Hampshire hog of America that is supposed to have its origin in Hampshire, England, are undoubtedly descendants from the herds of the Nobility in Holland. The development of these animals with this distinct belt of white between coal black ends of the body, is counted a most skillful work of breeding.

In size, the Dutch Belted cattle rank about with the Ayrshires. The general form and conformation is more like that of the Holstein. The cows weigh from 900 to 1300 pounds; the bulls weighing from 1600 to 2000 pounds. The dairy type is highly developed, being of a highly nervous temperament and very quiet disposition. Their most distinctive characteristic is the presence of the belt around the center of the body. This belt should extend around the body from just behind the shoulders to just in front of the hips. Otherwise, the body is coal black.

Importation and Distribution in America. — The first importation of Dutch Belted cattle into America, of which we have any record, was made by D. H. Haight in 1838. He made a second importation in 1848. The animals from these two importations were scattered throughout Orange County, N.Y., where their descendants are found at the present time. Robert W. Coleman also imported a large herd to place on his estate at Cornwall, Penn. The cattle of this breed in America to-day are nearly all descendants of these two herds. In 1840 P. T. Barnum imported a number of Dutch Belted cattle for show purposes, but soon sent the cattle to his farm in Orange County, N.Y. In 1906 one heifer was imported, but prior to this date none were imported for a period of over 50 years chiefly on account of the scarcity of the animals in their native land. Within recent years, importations have been made into Canada, Cuba, and Mexico.

In the United States, cattle of this breed are found most numerously in the Eastern states, especially New York, New

Jersey, Pennsylvania, Ohio, and the New England states; a few are found in the South and several small herds are to be found on the Pacific coast.

Dairy Characteristics. — The cattle of this breed are reputed to be excellent milkers. A few cows hold records of over 400 pounds of butter per year. At the Pan-American Exposition at Buffalo in 1901, five Dutch Belted cows were entered in the cow test in competition with the same number of cows of nine other breeds. While the five cows of this breed came out last in the test, they made a very creditable showing. The average production per cow for this breed was 4978 pounds of milk and 169.4 pounds of butter fat during the six months' trial. The average per cent of fat was 3.4. One breeder in New York, after keeping yearly records for eight years, reports twenty-five cows and heifer averaged 9000 pounds of milk for one year. A New Hampshire breeder reports eleven cows in his herd, making an average of 8579 pounds of milk in one year. One cow in this herd produced 12,672 pounds of milk in one year and 60,297 pounds in a period of six years with an average butter production during the same length of time of 596 pounds.

Herd Association. — The Dutch Belted Association which was organized in America in February, 1886, has registered between 2000 and 3000 animals. In Holland, the cattle of this breed are registered in the Netherland General Staurbuck, published at the Hague.

FRENCH–CANADIAN

This breed is descended from the native cattle of the province of Normandy and Brittany, France, and is very closely related to the Guernsey and Jersey breed on this account.

The French settlers who came from these French provinces and settled in Quebec, Canada, brought their cattle with them and have bred and improved them under the existing conditions and environment for more than 250 years. They have become adapted to severe climates and are noted for their vigor and ability to withstand the cold winters of the North.

In size and conformation they are very much like the Jerseys and Guernseys, the cows averaging about 700 to 900 pounds at maturity. They are active and well adapted to hilly and rough pastures. In color, they are black or black with a fawn or orange-colored stripe down the back and around the muzzle. In quantity of milk they compare with the Jersey, and the fat content averages between 4 and 5 per cent. Five cows of this breed were entered in the Dairy Demonstration at the Pan-American Exposition at Buffalo in 1901 and they stood seventh among the ten breeds represented on the basis of net profit in production. The five cows averaged 4932.8 pounds of milk and 196.2 pounds of fat for the six months.

The French-Canadian Herd Book was established in Canada in 1886. Cattle of this breed are found mostly in the Province of Quebec, but also in small numbers in other parts of Canada and in New York and the New England states.

POLLED JERSEYS

This breed originated in Ohio and is the result of the effort on the part of a few breeders to develop an animal with the characteristics of the Jersey, but without horns. In establishing this breed pure-bred Jerseys have been crossed with animals with alien blood that were hornless, and with this has been combined the blood of pure-bred Jerseys that were naturally polled. Animals of this breed have the same characteristics as their horned ancestors and have won attention on their merits. Nearly 1000 animals have been recorded.

The American Polled Jersey Herd Book Association was formed in 1895 with headquarters at Springfield, Ohio. The first volume of the Herd Book has not yet been published. Animals must be at least six months old and be born polled to be eligible to registration. A number of animals are now recorded in both the American Jersey Cattle Club and in the American Polled Jersey Herd Book.

Five animals of this breed were entered in the Dairy Demonstration held in connection with the Pan-American Exposition at Buffalo in 1901. The average production of the five cows for six months was 4065.6 pounds of milk and 190 pounds of butter fat, making an average test of 4.66 per cent.

KERRY CATTLE

The Kerry, the smallest of all the dairy breeds, is the native breed of cattle of Ireland where it originated and has been bred for centuries. Very little is known of the history of this breed prior to the middle of the eighteenth century. Pringle writing in 1872 described this breed as "A light, neat, active animal with fine and rather long limbs, narrow rump, fine small head, lively projecting eye, full of fire and animation, with a fine white cocked horn tipped with black, and in color either black or red." Relative to their size he described them as follows: "38 inches in height at the shoulders, 70 inches in girth, and 42 inches in length from the top of the shoulder to the tail head and a weight of about 30 'imperial stones.'" In regard to their products he states, "The average daily yield of milk of a Kerry cow properly fed and attended to is 3 gallons — a quantity capable of producing 6 to 7 pounds of butter weekly." Another type of the Kerry known as the Dexter Kerry originated by the introduction of some foreign blood; it is claimed by some that they are a result of a cross of the Kerry and Devon cattle of England. They differ from the original Kerry by being somewhat larger, very low set, round body, full hind quarters, and with short, thick legs, a heavy head, and rather straight horns.

There is very little difference in the two breeds of Kerries to-day; the points desired in either type are much the same; about the only difference is in the fineness of bone and refinement about the head in the true Kerry. The average cow weighs between 600 and 700 pounds. A black color is preferable, although other colors are often found. The most common color is black with a little white on udder and under the belly. They are vigorous and good rustlers, but are slow in maturing.

There has been a number of animals imported into this country, but they have not been widely distributed over the United States. Several herds are to be found in the Eastern states. The Kerry is best adapted as a family cow. They give a fair amount of milk containing about 4 per cent of fat.

CHAPTER IX

DUAL-PURPOSE CATTLE

GENERAL DISCUSSION — SHORTHORNS, RED POLLS, MINOR BREEDS

Definition of the Term. — The term "dual-purpose" is used to describe those breeds of cattle which are bred for both milk and beef production in contrast with those called special purpose which are bred primarily for either milk or beef. However, the question of beef production in connection with that of milk is largely one of degree since practically all dairy cattle are used for beef when their period of usefulness as milk producers is at an end.

Much of the discussion regarding the question of dual-purpose as compared with special-purpose cattle comes from erroneous ideas of what constitutes a dual-purpose animal. The man who is interested primarily in milk production is inclined to call every cow that does not produce milk profitably, especially if she shows a tendency toward beefiness, a dual-purpose cow. The cow in question might more often be correctly classed as a no-purpose cow. Other dairymen go to the other extreme, and call such cows as the Holstein dual-purpose because they have some value for beef production.

The breeder of pure-bred Shorthorns often calls any cow showing a good-sized udder when fresh a dual-purpose cow.

He is inclined to minimize the dairy part of the dual-purpose, as a rule, but at the same time points to large records which have been made by animals of the breed as proof that the breed has dairy qualities.

The true dual-purpose type stands about midway between the extremes of the dairy type, or large milk producers, and the beef type with little tendency toward milk production. A dual-purpose cow is one that produces a medium quantity of milk for a dairy cow, and that will sell at a fair price for a beef animal when fattened. A dual-purpose breed is one in which these characteristics are fixed so they are transmitted with reasonable certainty. There is occasionally a dual-purpose Angus or Hereford cow, but these breeds cannot be so classed, for the reason that this characteristic is not transmitted. Again we find cows in the dual-purpose breeds that are such remarkable milkers and show such inferior beef-making characteristics that they should be classed individually as special dairy animals, and not as dual-purpose. There is such a thing as a dual-purpose cow, if we correctly define the limitations of the term. It must not be expected that a cow of this type will compare as a dairy animal with good individuals of the special dairy breeds, or that her calves will be able to compete in beef production with those of the special beef breeds. A dual-purpose cow should be expected to produce about 200 pounds butter fat per year against about 300 for an equally good specimen of the dairy breed, and her calves should make fair beef animals.

Adaptation of the Dual-Purpose Cow. — The question whether the farmer should breed special-purpose dairy cattle or dual-purpose when dairy products are sold from the farm is one that has called forth endless discussion. The view represented by one extreme is that if a cow is to be milked at all, she must be of a special dairy breed, and that no such thing can exist as a profitable milk and beef producer combined. The other extreme holds to the view that the average farmer, who produces the largest bulk of the dairy products of the country, can make the best use of a cow that will produce a fair amount of milk and at the same time raise a calf which will be salable for beef purposes.

The question can best be considered by eliminating the points upon which there is practically no difference of opinion and concentrating attention upon the points where there is a chance for difference of opinion.

(1) It is generally admitted by all that cows of the special dairy breeds will, on the average, produce milk and butter cheaper than those of dual-purpose type, and that the special beef breeds excel the dual-purpose in beef production.

(2) The man who intends making dairying his chief business, with everything else of secondary order, should make use of the special dairy cow, and the man who produces beef animals and does not milk the cows should make use of the special beef breeds.

(3) The highest development of both milk and beef production cannot be combined in the same animal.

This leaves the general farmer for whom the dual-purpose cow is adapted, if for anybody. This large class of farmers, especially in the Central States, sells a number of things from the farm, among which milk or cream occupies a more or less important position.

The main question regarding the dual-purpose cow is whether this type is better adapted for this type of farmer than the special dairy breed. If a farmer of this class, who is using dual-purpose cattle, is asked why he does not use special-purpose dairy cattle in preference to the dual-purpose, he will give one or all of the following reasons: —

(1) The calves from the dairy breeds are not salable for beef purposes.

(2) The cows of the dairy breeds are not salable for beef when they are no longer useful for milk production.

(3) The cows of the dairy breeds are delicate, and require better care and attention than he can give them.

There is some ground for these statements. The calves of the dairy breeds, as a rule, cannot be raised for beef with profit, and it is often unprofitable even to raise them for veal.

It is also true that cows of the dairy breeds are not readily salable for beef when no longer profitable in the dairy, and when herds are properly weeded out it is necessary to dispose of a considerable number of cows every year. On the other hand, profitable dairy cows when once secured usually are kept at a profit for a comparatively long time, and the difference in return for the milk produced by a cow of the dairy breed and of the dual-purpose breed far more than makes up for the difference in beef value when placed upon the market. It is also true that cows of the dairy breed need good attention, or they will not be kept at a profit, but it cannot be said that they are especially delicate, although they will be of little value if allowed to go without proper shelter and food. The dual-purpose cow that produces milk for a few months during the summer only is as well adapted as the highly developed dairy cow for the farmer who will not provide the proper conditions. It requires good intelligent care to make use of as highly developed an animal as the modern dairy cow, and the farmers who are not able to meet these requirements might as well let her alone.

There is more to recommend the dual-purpose cow in the corn belt than elsewhere, and here the type is the most numerous. The typical farmer of this region finds it impracticable to secure sufficient labor to carry on a herd of dairy cows large enough to consume the feed, especially the roughness, grown on the farm. The raising of a number of beef cattle allows this surplus roughness which cannot be put upon the market with advantage to be utilized with small additional labor.

The dual-purpose cow also serves a useful purpose in many cases as an intermediate step in changing from a system of beef production to milk production when conditions make this change necessary. That is, when the farmer who has been engaged in beef production begins to sell dairy products, he usually milks the cows he has for a time, and gradually changes toward a dairy type by using dairy-bred sires. In this way he gains experience in handling dairy cattle gradually as the herd is developed.

Fully as many difficulties are experienced in breeding dual-purpose cattle as in breeding special-purpose dairy cattle. One of the tendencies observed is for individual breeders to emphasize either the beef or the milk production side, instead of keeping the two of about equal importance. This results in dual-purpose breeds varying much in type as bred by different breeders. The judging of dual-purpose cattle in the show ring is often unsatisfactory on account of the lack of a definite standard for such types, and the tendency of many judges is to minimize either the beef or milk-producing characteristics.

THE SHORTHORN BREED

Origin and Development. — This breed takes its name from its characteristic short horns. It is also known in some localities as the Durham breed, from one of the counties in which it originated. The original home of the breed is in Northeastern England, in the counties of Durham, Yorkshire, and Northumberland, especially in the valley of the River Tees. In this region the breed was improved and developed, and from here it has spread over almost the entire civilized world.

The exact origin is veiled in obscurity. The Romans, Saxons, Danes, and Normans brought their cattle in succession to England and mixed them with the native stock. After the invasion of the Normans there was little interchange of cattle for several centuries, and during this time the animals in the rich valley of the Tees probably increased in size, due to the favorable conditions of climate and food.

It is known that a large type of cattle existed in this region several centuries before the development of the modern Shorthorn in the eighteenth century. Early in the eighteenth century there were two general types, known as the Tees-water and the Holderness. About this time it is believed by most authorities, but disputed by others, that bulls were brought from Holland and used in some of the herds from which descended the improved Shorthorns.

The beginning of the improvement which resulted in the modern Shorthorn began about 1780, when Robert and Charles Colling began their breeding operations, which lasted until 1818. They are often spoken of as the founders of the modern Shorthorn breed. Shorthorns as bred by the Colling Brothers were generally good milkers, and this was considered by them as an important characteristic of the breed, to be retained as far as possible. At the same time they were more interested in developing the general symmetry and beef-making characteristics. They followed the methods of Robert Blakewell closely, practicing in-and-in breeding constantly.

Toward the latter part of the eighteenth century, Thomas Bates began breeding Shorthorns. He aimed constantly to develop a superior dairy and beef animal combined, and succeeded to a marked extent. Most of the best milkers among the Shorthorns at present are descended from animals of this breeding.

The Booth family began breeding Shorthorns about 1790. They have all along emphasized the beef production, and paid little or no attention to the dairy qualities.

Amos Cruickshank, who began breeding in 1837, developed the so-called Scotch type of Shorthorns, which are characterized by superior beef qualities and decidedly inferior dairy qualities.

The original Shorthorns were counted good dairy animals. Some very creditable reports are given regarding daily or weekly production of certain cows in the time of the Collings. It would appear, however, that even then the breed had the characteristics of the Shorthorn of to-day; milking heavy for a short time, but lacking persistency.

The Shorthorn cows brought to America during the early importations were usually at least fair, and some were exceptional milkers. On account of the general use of this breed for exclusive beef production in America, the dairy qualities were generally neglected, and most breeders aimed only at the best beef animal. This condition was further strengthened by the importations of the Scotch type. As a result, the typical pure-bred Shorthorn as found to-day in America as a rule has no claim whatever to be called a dairy animal. In a few localities the original dairy qualities have been preserved, and it is still possible to find first-class dairy animals in this breed in these localities. Within recent years there is an apparent revival of interest in the milking qualities of this breed, and a number of herds of pure-bred Shorthorns are now to be found where all cows are milked and records of individual production kept.

Form and Characteristics. — The Shorthorns vary in type from the extreme beef conformation to the dual-purpose, with a few of real dairy form. The latter are exceptional, and not typical of the breed in America. Shorthorn cows of the milking types weigh usually between 1200 and 1350 when mature. A typical cow of this type loses considerable flesh when in milk; when dry they fatten rapidly, and show much more of the beef characteristics.

Red, white, and roan are the typical Shorthorn colors. In disposition they are quiet and gentle. The calves of this breed weigh from 70 to 90 pounds at birth, and are strong and vigorous. In size and vigor the calves are exceeded by the Holstein and Brown Swiss breeds only among those commonly used for milk.

Dairy Characteristics. — While the Shorthorn breed is not counted among the dairy breeds, still enormous numbers of grades of this breed are milked, especially in the butter-producing states of the Mississippi Valley.

The available yearly records of pure-bred Shorthorns owned by American experiment stations show the following averages: —

	No. Cows Represented
Pounds of milk per year . . . 6017	37
Average per cent of fat . . . 3.63	40
Pounds fat per year 218	37
Average per cent total solids . 12.85	3

The above figures are about typical also of those herds in the hands of private breeders where some attention has been paid to the selection of the individual animal, and where fairly good conditions are maintained. It is considerably higher than the yield realized on the ordinary farm. At the St. Louis World's Fair in 1904 the Shorthorns competed against the Jerseys, Holsteins, and Brown Swiss. Twenty-four cows averaged 4152 pounds of milk and 153.4 pounds of fat in 120 days, and the leading Shorthorn in both milk and fat production produced 5207 pounds milk and 208.4 pounds fat in 120 days.

Some very creditable records of individual animals have been made by breeders that pay attention to the dairy qualities. The private records of a herd in Pennsylvania show an average for 38 cows of 9031 pounds of milk per year and of 8515 pounds of milk for 52 cows and heifers.

The cow Mamie Clay 2d in this herd produced 47,048 pounds of milk with her first four milking periods.

The following are the best records available for pure-bred animals of this breed in America at present: —

Cow	Owner	Milk	Butter Fat
Rose of Glenside . .	May & Otis	18,075	625
Lula	U. of Missouri	12,341	515
Panama Lady . .	U. of Missouri	13,789	490
Florence	U. of Nebraska	10,438	424
College Moore . .	Ia. Agric. College	9,896	407

The milk of the Shorthorn cow contains on the average about 12.5 per cent of total solids, of which from 3.60 to 4 per cent are fat. The milk in color and in size of fat globules ranks next to the Channel Island breeds, and between them and the Holsteins and Ayrshires.

Shorthorns in England. — The dairy Shorthorn is the principal dairy cow of England to-day and the typical Shorthorn cow of England as well. Professor Long says: "The milk-producing farmer has studied how to increase the flow of milk while maintaining the characteristic feeding qualities of the breed, and has succeeded. On the other hand, the great pedigree breeders have subordinated milk to flesh development, form, quality, and even color. Where sires and dams are of equally renowned milking character, the Shorthorn is preëminently the best dairy cow in the best dairy country in the world."

The milking qualities of the English Shorthorn are shown in a remarkable way by the results of the milk and butter tests made at the London Dairy Show. In the eleven years from 1894–1904 inclusive, first place in both milk production and fat production was won in every case by a Shorthorn competing against Jerseys, Guernseys, Ayrshires, Red Polls, and Crosses. In seven of these tests the winner was a non-pedigreed Shorthorn, and in four a registered Shorthorn. Two hundred and thirty-six pedigreed Shorthorns in these eleven tests averaged 48 pounds milk and 1.82 pounds fat per day, with an average of 3.88 per cent fat. One hundred and twenty-one non-pedigreed Shorthorns averaged 51.8 pounds milk and 1.95 pounds fat daily, with an average of 3.79 per cent of fat.

Breed Organizations. — The registration of pure-bred animals of this breed is in the hands of the American Shorthorn Breeders' Association, with headquarters in Chicago. There is no provision made for making official tests or recording dairy records of this breed. The Dairy Shorthorn Breeders' Association was organized in 1910.

RED POLLS

This breed originated in the counties of Norfolk and Suffolk, England. These are adjoining counties in the eastern part of England. They are low and flat, with some marsh land. The soil is naturally diversified, although rather poor, but has been brought to a high state of fertility by good management. About 80 per cent of the area is tillable. The climate is generally typical of England, although the rainfall, which averages twenty-six inches per year, is less than the average for England.

The origin of this breed, like that of most others, is uncertain. It seems, from the best information obtainable, that they have been bred up from the cattle which have been native to these two counties since a time farther back than accurate historical records go. The beginning of improvement began in the latter part of the eighteenth century as part of the widespread movement of that time for improved livestock.

Mr. Euren, the first Secretary of the Red Polled Herd Book, is of the opinion that the original cattle of this type were brought to England by the Danes who settled in this part of England in the fifth century. He does not believe there is any foundation for the statement often made by writers that Galloway blood was introduced from Scotland by bringing Galloway bulls. However, Youatt, the well-known early English writer on cattle, and Wallace, a recent writer, attribute the improvement largely to this source.

In 1804 Young wrote an account of the agricultural conditions in Norfolk and Suffolk, in which he describes the cattle of Norfolk as a small, red class of cattle, partly polled and partly horned. They were bred at this time more for beef than for milk, and made a poor impression on the writer. Later the horns were bred off by using probably either Suffolk or Galloway bulls for this purpose. The beef qualities were considerably developed, according to some authorities, by introducing Devon blood. In 1818 the cattle of Norfolk began to be known as Norfolk Polled.

The cattle of Suffolk from the earliest records were known as remarkable milkers. They were rather small in size, red, brindled, dun or mouse colored, and always polled. Young, writing in 1804, describes them as follows: " A clean throat with little dewlap, a clean head, thin legs, a very large barrel, ribs tolerably springing from center of the back, but with a heavy belly, backbone ridged, chin thin and hollow, loin narrow, udder large, loose and creased when empty, milk veins remarkably large. A general habit of leanness, hip bones high and ill covered with flesh."

According to the same writer yields of five gallons of milk per day were not uncommon for entire herds while on pasture. The breed had the reputation of being the heaviest milkers in England for their size, and of being especially adapted for poor pastures and unfavorable surroundings. The Suffolk cattle were always bred mostly for milk. The cattle of this county became known as Suffolk Polled. The colors other than red were bred out during the early part of the nineteenth century.

The cattle of Suffolk and Norfolk were developed along the same lines, until about 1846 it was generally recognized that the two types were so near together that they were practically the same. The division in name continued until 1862, when at the Royal Agricultural Society Show they were first classed together under the name, Norfolk and Suffolk Polled. In 1882 the name was shortened to the present name of Red Polled.

At the present time the breed is found mostly in the two counties where it originated, and to some extent in Australia and New Zealand. They are not found on the Continent of Europe. They are used in England as a dual-purpose cattle, and rank in that class about with the non-pedigreed Shorthorns.

Importation and Distribution in America. — Cattle from the home of the Red Polled breed were undoubtedly brought to America during the colonial days, but were not kept pure. These early importations are probably responsible for the muley red native cows often seen, especially in the Eastern states, some years ago. The first importation of the improved type was brought to America in 1873 by Gilbert F. Tabor of Patterson, N.Y., who also made several later importations. Several importations were made between 1880 and 1890, and from these mentioned are descended most of the cattle of this breed now in America.

They are now found in all or nearly all the states of the Union, but are the most numerous in Ohio, Illinois, Michigan, Iowa, Wisconsin, and Kansas. The English Red Polled Herd Book was first issued in 1874. The Red Polled Cattle Club of America organized in 1883 registers the breed for America. There have been close to 30,000 animals registered up to 1910.

They have won their way entirely upon their merits, and are increasing rapidly in those states where dual-purpose cattle are in demand. They are the most typical and most popular of the real dual-purpose breeds.

Form and Characteristics. — In size the Red Polls rank below the Shorthorn and other heavy beef breeds. The cows weigh between 1200 and 1300 as a rule, but occasionally one reaches 1500 or more. The bulls range from 1800 to 2200, with occasional individuals reaching 2500 at maturity.

In color the Red Polls are a deep cherry red. White may appear on the tip of the tail, the udder may be white, and a few small white markings are allowed on the belly. White on any other part of the body disqualifies the animal, as does a black nose or even abortive horns.

In form the Red Polls are typical dual-purpose cattle. In general the conformation is about midway between the dairy and beef types, as is typical of the dual-purpose breeds. The general form is parallelogrammic. The head and neck are lean and of dairy type, with the characteristic poll. The hind quarters are of moderately good beef form. The udder is usually somewhat pendulous, and the fore quarters are often deficient and irregular in shape. The udder is seldom meaty in character, but usually elastic and mellow in quality. The milk veins and milk wells are usually fairly prominent. The teats are inclined to be irregular in shape, and usually large. In some animals the extreme size of the teats is somewhat objectionable.

There are no generally recognized distinct types of this breed at present, although different herds vary widely, depending upon the purpose for which they have been bred. In some cases the breed is bred only for beef production and the cows are not milked. In other herds the dairy qualities are given first place. This results in wide variations in form and capacity for milk production in different herds. The judging of Red Polls in the show ring has usually been done mostly from the beef standpoint, which tends to develop the beef qualities at the expense of the milking qualities. The Red Polled Cattle Club has recently formulated a scale of points to assist in establishing a more definite type.

Red Polls are fair breeders, ranking about with the Shorthorns in this respect. In regard to early maturing qualities, they rank probably rather behind the Shorthorns.

As Beef Producers. — This breed ranks well as beef producers, but does not win highest honors in competition as beef animals. The steers are satisfactory feeders, gain rapidly, and bring a creditable price. The steers are not so blocky and compact as those of the strictly beef breeds, and usually are longer legged. The cows fatten rapidly when they are not producing milk.

Red Polls as Milk Producers. — This breed ranks high in this respect for a dual-purpose breed. But few records are available from herds owned by experiment stations. The following summary includes all available: —

		No. of Animals Represented
Pounds milk per year	5906	9
Average per cent fat	4.03	9
Average yield fat per year	238	9

In yield of milk the Red Polls are excelled by the Holstein and Ayrshires and in England by the Dairy Shorthorns, while they probably slightly lead the Jerseys and Guernseys.

As compared with the Shorthorn breed as found in America, the Red Polls are much superior as milkers on the average, while as regarding beef production the conditions are reversed. The Red Polls are a real dual-purpose breed according to the definition given, since they are quite uniform in their ability to produce a fair amount of milk and are salable at a creditable price for beef. Their milk is about the proper composition to make it most suitable for market purposes or cheese making. In color of milk and butter the Red Poll ranks about with the Shorthorn, below the Jersey and Guernsey and above the Holstein and Ayrshire.

At the Pan American Exposition Dairy Test in 1901, covering six months, five Red Poll cows stood fifth among ten breeds. The Red Polled cow, Mayflower 2d, stood second, among the fifty cows taking part in the test.

A system of keeping authentic records of milk and fat production was begun in 1908. The highest record for 1909 was by Gold Drop, 11,298, with a production of 11,889 pounds of milk and 511 pounds of fat. A herd owned in Minnesota averaged 6564 pounds of milk and 259 pounds of fat from 10 mature cows, which included all the herd. Other private records of merit are: Neva N., 12,204 pounds of milk and 469 pounds of fat; Mayflower 2d, 10,458 pounds of milk and 469 pounds of fat in one year.

SCORE CARD — RED POLLED COW

DISQUALIFICATIONS — *Scurs, or any evidence whatever of a horny growth on the head. Any white spots on body above lower line or brush of tail.*

Color — Any shade of red. The switch of tail and udder may be white, with some white running forward to the navel. Nose of a clear flesh color. Interior of ears should be of a yellowish, waxy color 2
    Objections — An extreme dark, or an extreme light red is not desirable. A cloudy nose or one with dark spots.

Head — Of medium length, wide between the eyes, sloping gradually from above eyes to poll. The poll well defined and prominent, with a sharp dip behind it in center of head. Ears of medium size and well carried. Eyes prominent; face well dished between the eyes. Muzzle wide with large nostrils 6

Objections — A rounding or flat appearance of the poll. Head too long and narrow.

Neck — Of medium length, clean cut, and straight from head to top of shoulder, with inclination to arch when fattened, and may show folds of loose skin underneath when in milking form 3

Shoulder — Of medium thickness and smoothly laid, coming up level with line of back 6

Objections — Shoulder too prominent, giving the appearance of weakness in heart girth. Shoulder protruding above line of back.

Chest — Broad and deep, insuring constitution. Brisket prominent and coming well forward 10

Back and Ribs — Back medium long, straight and level from withers to the setting on of tail; moderately wide, with spring of ribs starting from the backbone, giving a rounding appearance, with ribs flat and fairly wide apart 14

Objections — Front ribs too straight, causing depression back of shoulders. Drop in back or loin below the top line.

Hips — Wide, rounding over the hooks, and well covered . . 3

Quarters — Of good length, full, rounding, and level; thighs wide, roomy, and not too meaty 6

Objections — Prominent hooks, sunken quarters.

Tail — Tail head strong and setting well forward, long and tapering to a full switch 2

Legs — Short, straight, squarely placed, medium bone . . . 3

Objections — Hocks crooked, legs placed too close together.

Fore Udder — Full and flexible, reaching well forward, extending down level with hind udder 10

Hind Udder — Full and well up behind 10

Teats — Well placed, wide apart and of reasonably good size . 4

Objections — Lack of development, especially in forward udder. Udder too deep, "bottle shaped," and teats too close together. Teats unevenly placed and either too large or too small.

Milk Veins — Of medium size, full, flexible, extending well forward, well retained within the body; milk wells of medium size 6

Hide — Loose, mellow, flexible, inclined to thickness, with a good full coat of soft hair 5

Objections — Thin papery skin, or wiry hair.

Condition — Healthy; moderate to liberal flesh evenly laid on; glossy coat; animal presented in good bloom . . . 10

   Total 100

General Description — Cow medium wedge form, low set, top and bottom lines straight except at flank, weight 1300 lb. to 1500 lb. when mature and finished.

POLLED DURHAMS

This breed originated in Ohio, and has the distinction of having been developed in America. In reality they are Polled Shorthorns, since Shorthorn blood was used almost exclusively in their development. A portion of the Polled Durhams are pure-bred Shorthorns that are descended from natural polled animals, and this character was fixed by proper selection and breeding. These are now known as double-standard animals. The other branch was formed by breeding the "native" or "muley" cows to pure-bred Shorthorn bulls. Most of the Polled Durhams are descended from herds formed in this way. They are found most numerously in Ohio, Indiana, Illinois, and other Central and Middle Western states at the present time.

The American Polled Durham Breeders' Association was formed in 1889, and the first volume of the herd book was published in 1894. To be eligible to entry in the herd book the animals must carry at least $96\frac{7}{8}$ per cent Shorthorn blood and have the color markings and general conformation of the Shorthorn and be entirely without horns. Some animals of this breed are now registered in both the Shorthorn Herd Book and the American Polled Durham Herd Book.

The breeders have given considerable attention to the development of the milking qualities, and as a breed, they excel the Shorthorn in this respect. The description given of the conformation, size, and characteristics of the Shorthorns will answer for a description of this breed, as they are essentially alike, except that the Polled Durham is hornless. The hornless feature has won popularity for this breed. A number of animals of this breed have been exported into South America and other countries.

DEVONS

The Devon cattle are commonly classed as dual-purpose cattle, and are one of the oldest breeds that originated in Great Britain. Very little is known of the early history of this breed, but it is generally believed that they are related to the Hereford and Sussex breeds. Some writers believe that the above-named breeds descended from the Devons. These cattle were bred and developed in Devonshire, England, and are divided into the North and South Devons. The North Devons more nearly represent the original and true type of this breed. The Quartly family of North Devon was largely responsible for the development and improvement made with the Devons; Francis Quartly, particularly, took the lead in the improvement of this breed, which made them popular.

While the breed are classed as dual-purpose animals to-day, they originally were bred for milk, beef, and draft purposes in their native country.

The Devons were first imported into America in 1817. Several importations have been made since this date, and they are popular in some localities in the United States, especially where the production of milk is not the sole object for which the cattle are kept. Representatives of this breed are found in nearly every state in the Union. The cows average about 1000 pounds in live weight, and are hardy, strong, and active. In color, the Devon is a deep red. White markings may appear on the belly and on the udder. The horns are long, and turn upwards and backwards.

The dairy characteristics of this breed have never been very highly developed; like other dual-purpose animals, they give a large quantity of milk at the beginning of their lactation period, but are not persistent milkers. In a report made by the New York Experiment Station, the average butter fat test for 72 head of cattle was 4.15 per cent. Yields of 5000 pounds of milk per year have been reported. The interests of this breed are looked after in America by the American Devon Herd Book Association. The first volume of the Herd Book was published in 1861.

CHAPTER X

STARTING A DAIRY HERD

SELECTION OF BREED — COMMUNITY BREEDING — PRECAUTION AGAINST DISEASE

THE first question that arises in starting a dairy herd is the question of breed. The tendency is to attach too much importance to this matter, as influencing the success of the venture. As already emphasized, the importance of making use of a pure breed of some kind is more apt to be under than overestimated. As a rule a man that will make a success with one breed will be about equally successful with another. The characteristics of the several breeds have been discussed in a previous chapter, and a study of these descriptions will enable a choice to be made.

Some consideration must be given to the purpose for which the product is to be used. For example, if milk is to be sold in a market where richness is not recognized, breeds giving the larger yield would naturally be the choice. If butter or cream is the product to be marketed, other factors need to be considered.

It is impossible to give data that are satisfactory regarding the relative production of the breeds. The best figures the author has been able to gather are brought together in the table which follows. These are yearly records of pure-bred animals as reported by experiment stations in the United States for animals owned by them:—

	Average Pounds Milk per Year		Average per cent Fat for Year	
	No. Cows	Lb. Milk	No. Cows	Per cent Fat
Holsteins	83	8699	83	3.45
Jerseys	153	5508	154	5.14
Shorthorns	37	6017	40	3.63
Red Polls	9	5906	9	4.03
Guernseys	17	5509	21	4.98
Ayrshires	24	6533	24	3.85

It is assumed that the conditions under which these records are made are fairly comparable, although certainly no more favorable than are found in good herds owned by individuals. The Shorthorn records are from cows selected especially for milking characteristics, and in many cases other cows of this breed in the same herd are not milked at all.

Hoard's *Dairyman* during the past eight years has published statistics regarding the amount of milk and fat sold at creameries and cheese factories in various states, under the title of "Cow Census." The following compilation of results is made by the author from these data published. Only records are included from herds, the animals of which are designated under a breed name without any qualification. All described as "grade" or "mixed" are rejected. While these figures leave much to be desired regarding accuracy, they serve to give a general idea of the amount of milk and fat actually marketed per cow by the producer. A portion of the results are reported for milk yield only and others for fat.

	No. Herds	No. Cows	Average Lb. Milk	Average Lb. Fat
Shorthorns	48	468	3708	
	115	1139		146
Jerseys	83	967	3816	
	123	1224		201
Holsteins	45	647	5294	
	34	434		185
Ayrshires	9	175	3801	
	5	114		191
Guernseys	4	55	4961	
	8	102		238

The figures given represent the amount sold, not the amount produced. Since most of the herds are small, the product consumed by the family of the producer would add to the figures. In some cases probably certain herds did not supply milk to the factory during the entire year. All breeds are on the same basis, however, and where the number is sufficiently large, probably give a fair idea of the amount of product sold per animal by the average producer. This astonishingly low production is due largely to the lack of a proper selection of individuals, and partly to improper feeding and management.

Community Breeding. — Before making a choice as to the breed of cattle to be developed on any farm, it is well to consider the matter from the community standpoint. Instead of selecting a breed because it is not well represented in the community, as is sometimes done, it is much better to develop a herd of the same breed as already predominates in that locality. The importance of this community breeding is now recognized, and systematic efforts are being made in various states to promote local organizations for the purpose of furthering the movement. When a large number of well-bred animals of the same breed is found in one locality, a community organization is also possible which creates a new interest in the subject of breeding, and stimulates and educates the members to take advantage of their opportunities.

When official testing is done, the expense is greatly reduced by several members of the association having the testing done at the same time and thus dividing the traveling expenses. Many other advantages follow active work by such an organization, such as buying cattle or feed together when desirable, advertising the stock for sale, and combating disease. As a rule, such organizations include only those interested in a single breed and within a radius where it is possible for all to coöperate and attend the meetings.

Where one community has a large number of animals of the same breed, the fact soon becomes widely known, and a good market for that class of animals is established. Buyers are always attracted by the possibility of buying a number of cattle in one neighborhood, and surplus stock can be disposed of much more readily than where the animals are widely scattered. Certain localities become known as Holstein centers, others as Jersey, or Guernsey, and buyers from a long distance visit these localities, knowing they will be able to find what they wish to purchase.

It is a great advantage also in regard to breeding animals. By trading bulls it becomes possible to retain the best bulls in service, and not only reduce the expense for this purpose, but also to make wide use of a bull that is found to sire especially valuable animals.

Starting a Dairy Herd. — The general plan to be followed in building up a dairy herd will naturally depend upon conditions, such as purpose in view, knowledge of the business, resources at hand, etc. As a rule, the production of dairy products and not of breeding stock should be made the foundation of the business. There is no objection or reason why the two should not be combined, but only in exceptional cases can a man expect to run a herd as a profitable business on the proceeds of sales of breeding stock alone.

If the object is the sale of dairy products alone, high-grade animals serve the purpose equally as well as the registered. Every grade herd, however, should have a registered bull of good individual merit and backed by good dairy records. The use of a good bull and careful selection of individual cows by keeping records will make it possible to build up a high-producing herd within a few years.

There are in general two ways of getting a good herd of dairy cows together. The first is by purchase, the second by breeding them. The first is the quicker, but can be followed only by those having ample capital. As a rule, good dairy cows cannot be had except at high prices. It is always the poor milkers that are for sale, and not the good ones. To get a herd of good milkers together by purchase is possible, but expensive. As a rule, where cows are all purchased and none bred, the average production is low and it never gets much better as long as this course is pursued. In addition to the certain result of finding that many of the purchased cows are unprofitable, there is great and constant danger of bringing disease into the herd with purchased cows. Tuberculosis and contagious abortion are often brought into a herd in this manner.

For the majority the best plan is to breed most of the cows to be used in the herd. If the herd must be started with the minimum outlay of capital, the best plan is to begin with the best cows that can be purchased in the locality; the herd bull should be selected with great care, and the poorer cows disposed of as soon as discovered, and replaced with heifers from the best cows. Within a few years a good herd is on hand with a small investment, and the owner has gained experience as the herd develops that will enable him to handle a high-producing herd when he has one. A man who has had no experience with cows, or whose experience has been limited to beef cattle or ordinary cattle of low dairy quality, is almost certain to make a failure, if placed at once in charge of a herd of dairy cows of high milk-producing capacity. A combination of the two plans is also very good. That is to buy most of the herd of such common cows as have indications of milking qualities, and to include a few choice cows and save their offspring most carefully.

The two things to be most emphasized in building up the herd are the selection of the herd bull and the elimination of unprofitable cows by keeping records of production by each individual.

A few with special liking and adaptation for cattle breeding can make a good success breeding registered dairy cattle for the purpose of selling the surplus as breeders. Unless one has plenty of financial backing, the sale of breeding stock had better be made secondary to the sale of dairy products. To carry on a high-class herd of registered dairy cattle that will produce offspring salable at good prices for breeding purposes requires unusual judgment in the selection of breeding animals and the keeping of records of milk and fat production. In addition, business ability of a high order is necessary to make a marked success along this line.

Crossing Breeds. — Crossing distinct breeds defeats the very object for which breeds have been developed. Breeds have been developed and kept pure in order that certain characters might become fixed so strongly that they will be transmitted regularly. Crossing breaks the chain of inheritance, and makes it impossible to predict what will be the outcome. As a rule, little is gained, and the outcome often is very disastrous. However, it is a very common practice with many. A farmer having perhaps a good grade herd of Jerseys observes the much larger yield of milk secured by his neighbor who breeds Holsteins and decides to make a Holstein cross, thinking he will combine the quality of the Jersey with the quantity of the Holstein. Occasionally this end is partially attained, but just as often the animal inherits the quantity of the Jersey and the quality of the Holstein. The next year the farmer possibly decides his animals are too small, and uses a Shorthorn to increase the size. The result of such practice is to lose the breed characters, and the occasional good animal that appears from such a mixture does not transmit any definite characteristics.

It is a well-known fact that as a rule the first cross between distinct breeds is good, and some animals may have the good characteristics of both breeds to some extent. Many inferior animals appear in the second generation, making the results of crossing unsatisfactory. The proper course to pursue is to first select the breed, after due consideration, that seems to meet the requirements or tastes of the breeder, then select the best individuals and the ones most likely to transmit these characters from this breed, and continue along the same line unless it is found after sufficient trial that a serious mistake has been made.

Guarding against Disease. — The only diseases that are likely to be brought into a herd by the purchase of animals are tuberculosis and contagious abortion, and these should be guarded against with the greatest care.

Tuberculosis does not develop spontaneously. It spreads only by the transmission of the germs which cause it from an infected animal. A cow may be badly affected with the disease without showing any evidence of the fact. The placing of such an animal with a healthy herd may result in the entire number contracting the disease within a few months. Fortunately the tuberculin test makes it possible to determine accurately whether an animal is infected. In starting a dairy herd every animal should be subjected to the tuberculin test by a competent person, and rejected unless sound. If this is done, and every animal that is added to the herd later is tested before being placed with the others, this dangerous disease may be excluded indefinitely. Failure to observe this precaution has resulted in many cases in the infection and loss of valuable herds.

It is a more difficult matter to guard against contagious abortion, and the financial losses resulting from it are severe. There is no means of testing a cow to learn if she carries the disease. All that can be done is to make careful inquiry regarding the health of the herd from which it is desired to purchase stock. It should be understood that the bull as well as the cows may carry the germs of this disease.

As a rule it is much safer to purchase stock from a breeder than from a dealer, since in the latter case little or no information is usually available regarding the dairy qualities of the cows or as to the health conditions in the herds from which the animals originally came. Further, animals in the hands of dealers have greater opportunity to become infected with disease from contact with others from many herds. The safest animals to purchase are the surplus from the herds of reliable breeders or from a good herd that is being dispersed.

CHAPTER XI

SELECTION OF THE INDIVIDUAL COW

BREEDS are of great value as a means of preserving and transmitting qualities which have already been developed, and it is highly important to select a breed adapted to the purpose for which it is to be used. However, the selection of the individual cow within the breed is of even more importance as effecting the economic production of milk.

The highly developed dairy cow of to-day is, to a large extent, artificial. While it cannot be said with certainty what the conditions were before cattle were domesticated, there is little doubt, judging from other species of animals still in the wild condition, that the cows produced only milk enough to support the calf for a few weeks until it could subsist on other food. There was probably little difference in the amount of milk produced by different individuals at this time, and the milking characteristic was undoubtedly transmitted quite uniformly.

After cattle were in a state of domestication and the milk became an important article of human food, some attention began to be paid to developing the milking functions. Through natural variations, certain animals showed more highly developed milk-producing functions than others, and by using these for breeding purposes, and through the stimulation of the mammary glands by better feed and regular milking, a change was gradually made in the amount of milk secreted, and probably to some extent in the quality as well. While the wild cow possibly produced from one to two thousand pounds of milk in a year, a good dairy cow of to-day may produce more than this in a single month.

It is a well-known fact and one easy to understand that when any characteristic or function has been developed to a higher degree in a breed of animals than existed originally, the acquired characteristics may not be transmitted regularly. There is a constant tendency for the characteristics of some of the ancestors to appear. The farther the animal is developed from the original form, the more difficult, as a rule, is the fixing of the acquired characteristics. This explains why there is such a wide variation in the capacity of individual cows to produce milk. While the wild cows presumably varied but little in the milk produced, it is not uncommon under present conditions for one cow to produce four or five times as much, or even more, than another individual of the same breed kept under similar conditions in the same herd. These individual variations must be expected, and the higher the development of the dairy cow is carried, the more difficulty will be encountered in keeping up to the desired standard.

The tendency in any race or breed of animals is to transmit characteristics normal to that breed or race. This explains the common observation that only a few of the progeny of a cow of unusual dairy qualities are the equal of the dam, and her offspring may be only ordinary for the breed.

While the importance of the selection of the individual cow has been recognized for a long time by those who have given the subject thought, it is only within recent years that the full significance of individual variation has been appreciated by those who are engaged in dairy farming. One of the weakest points in the system of dairy farming as carried on at the present time is the failure on the part of a large number to appreciate the importance of this factor of individual selection, or, if it is appreciated, a failure to give this subject the attention and time its importance justifies. Even where cows are milked regularly and the selling of dairy products is a regular business, the yearly butter production seldom exceeds 250 pounds per cow except in the hands of the special dairyman. Even among those who are making the sale of dairy products their principal source of income, the average production is comparatively low.

The Illinois Experiment Station,[1] after testing 18 herds, including 221 cows through complete milking periods, reported the average milk production to be 5617 pounds and fat production 227 pounds. The best herd averaged 350 pounds, the poorest 142 pounds of butter fat per cow. The ten best cows averaged 389 pounds, the ten poorest 142 pounds of butter fat per cow for the year. Herds which had been graded up by the use of a pure-bred sire produced 85 pounds of butter fat more per cow than did those in which no grading had been done. These herds were in the hands of men who were making the production of milk their principal business. As a result of this investigation it was concluded that at least one third of the cows in the ordinary herds which were being used for milk production in that state were unprofitable, and that on nearly every dairy farm a few cows were kept at an excellent profit, some at a small profit and some at an actual loss. A summary of these tests is given in Table 1: —

[1] Circular No. 102, Ill. Exp. Station.

TABLE 1

PRODUCTION OF AVERAGE BEST AND POOREST COW IN TWENTY-FOUR ILLINOIS HERDS

Herd	No. Cows in Herd	Pounds Milk			Per cent Fat			Lb. Butter Fat		
		Average	Best	Poorest	Average	Best	Poorest	Average	Best	Poorest
1	11	5753	6099	4391	4.54	5.17	3.91	262	315	172
2	8	7376	8739	4928	3.19	3.81	3.92	268	333	193
3	5	8057	9454	6719	3.42	3.40	3.27	276	324	221
4	11	6220	7445	4091	3.89	4.82	3.83	242	359	157
6	20	7873	9067	5796	3.62	4.41	3.65	285	399	212
7	10	4525	5507	3412	3.76	4.70	3.78	170	264	129
8	10	4486	6647	2691	4.29	3.09	3.61	193	263	97
10	13	5431	7291	3847	4.18	4.31	4.38	227	315	168
11	9	5969	6531	5552	3.43	3.78	3.01	205	247	168
12	13	4504	6429	2090	3.89	3.80	4.83	175	248	101
15	12	5128	6289	3491	4.03	4.74	3.01	207	299	135
16	9	4608	5293	3752	3.98	4.49	3.99	184	238	150
17	7	4355	6115	3710	3.96	3.31	3.33	173	203	124
19	19	5410	6413	4530	4.11	4.57	3.49	243	293	158
20	15	6106	7530	2980	3.84	3.93	4.56	235	296	136
21	15	5971	8882	4025	4.06	3.75	3.55	243	333	143
23	25	3314	4337	1846	4.28	4.96	4.24	142	216	78
24	9	5921	6911	3478	5.91	6.91	4.64	350	477	161
	221	5616	6994	3962	4.03	4.55	3.83	226	301	150

Table 2 gives the results of a year's test made by the Kansas Experiment Station [1] of a herd of fifteen cows which shows the same facts in a striking manner. The table is arranged in three divisions, the records of the best five cows are placed in the first group, the records of the next best five in the second group, the records of the poorest five in the third group. The

[1] Erf. Report Nat. Creamery Butter Makers' Association, 1906.

butter fat is counted at the low price of 15 cents a pound, and the cost of feed is also on a low basis, but this does not interfere with the comparison.

TABLE 2

Year's Record of a Herd tested by Kansas Experiment Station

FIRST LOT

Cow No.	Lb. Milk	Per cent Fat	Lb. Fat	Cost of Feed	Value Butter Fat	Value of Butter Fat over Feed	Cost of 1 Lb. of Fat
1	9.116	4.21	383.7	$32.80	$60.88	$40.37	$0.085
2	7.015	4.43	310.8	30.61	49.26	28.11	0.098
3	8.054	4.13	332.8	35.59	51.92	27.18	0.106
4	6.504	4.59	289.5	29.26	45.90	25.41	0.101
5	6.509	4.27	277.9	29.20	43.89	23.39	0.105
Av.	7.439	4.28	318.9	31.49	50.37	28.89	0.098

SECOND LOT

Cow No.	Lb. Milk	Per cent Fat	Lb. Fat	Cost of Feed	Value Butter Fat	Value of Butter Fat over Feed	Cost of 1 Lb. of Fat
6	5.742	3.48	199.8	29.55	31.02	9.22	0.147
7	4.772	3.92	187.0	27.25	29.88	8.27	0.145
8	3.475	5.14	178.6	25.24	28.16	7.60	0.141
9	3.913	4.14	161.9	27.27	25.41	3.41	0.168
10	4.200	3.96	166.3	27.69	25.38	3.28	0.166
Av.	4.420	4.04	178.7	27.40	27.81	6.35	0.153

THIRD LOT

Cow No.	Lb. Milk	Per cent Fat	Lb. Fat	Cost of Feed	Value Butter Fat	Value of Butter Fat over Feed	Cost of 1 Lb. of Fat
11	3.583	3.79	135.7	26.75	21.39	− 0.43[1]	0.197
12	2.903	4.13	119.9	22.89	18.11	− 0.87	0.190
13	3.730	4.23	157.8	31.22	24.34	− 1.86	0.198
14	2.141	4.74	101.5	24.43	15.30	− 6.25	0.240
15	3.089	4.06	128.7	26.32	19.78	− 2.35	0.204
Av.	3.089	4.19	128.7	26.32	15.98	− 2.82	0.206

[1] Value of food over butter fat.

It will be noted that the first lot were kept at an excellent profit, the average return for butter fat in excess of the cost of feed was $28.89 per animal and cost of fat per pound 9.8 cents. The second group were kept at an expense of only $4.09 less than the first group, while the value of their product on the average was $22.54 less. Each cow in this group produced only $6.35 worth of butter fat in excess of the cost of feed. The third group of five cows were kept at an expense of $5.17 each less than those in the first group. They produced such a small amount of fat that it lacked $2.82 of being equal to the value of the feed consumed. In this case, one third of the herd were entirely unprofitable. It is very evident that a larger net profit would have been made had the third lot been culled from the herd in the beginning. In addition to a loss of $2.82 each on the feed, there is to be considered the labor of caring for this third lot, the interest on the money invested, and the room occupied in the barn.

Table 3 gives the summary of twelve months' production for 719 cows owned in the Southern States, based upon records made by representatives of the Dairy Division, United States Department of Agriculture.[1] These data show the same wide variation in the cost of production by individual cows as found in the investigations previously noted. In this case the best cow produced butter at a cost of 13.4 cents per pound as compared with 37.1 for the least profitable. The best cow consumed feed worth $3.00 for each $1.00 expended for the poorest but at the same time she gave a return of $8.30 for each $1.00 by the inferior animal.

[1] Twenty-fifth Annual Report Bureau of Animal Industry, U. S. Dept. Agriculture, p. 67.

TABLE 3

Results of Twelve Months' Record for 719 Cows in the Southern States

Items	Average of 719 Cows	Best Cow	Poorest Cow	Average of Best 10 Cows	Average of Poorest 10 Cows	Average of Best 30 Cows	Average of Poorest 30 Cows
Milk produced, pounds	4299.40	8325.50	1125.00	8681.90	1577.60	7326.00	2099.60
Butter fat produced, pounds	216.84	538.79	64.12	459.00	77.21	391.75	100.70
Value of butter fat at 28c a pound	$60.71	$150.86	$17.95	$128.52	$21.62	$109.69	$28.20
Value of skim milk at 20c a hundredweight	8.17	15.57	2.12	16.45	3.00	13.87	4.00
Total value of products	68.88	166.43	20.07	144.97	24.62	123.56	32.20
Cost of feed per cow	36.27	72.03	23.80	65.73	24.63	54.83	27.36
Profit per cow	32.61	94.40	−3.73	79.24	−0.01	68.73	4.84
Cost of producing 1 lb. butter fat, in cents	16.70	13.40	37.10	14.30	31.90	14.00	27.20
Returns for each $1 invested in feed	$1.90	$2.31	$0.84	$2.20	$1.00	$2.25	$1.18
Profit on each $1 invested in feed	0.90	1.31	−0.16	1.20	0.00	1.25	0.18

This table shows the best thirty cows produced about three and one half times as much as the poorest thirty cows. The former gave a return of $2.25 for each dollar invested in feed to $1.18 for the latter. The best individual produced nearly eight times as much as the poorest.

Table 4 gives interesting data from the Storrs Experiment Station herd covering a period of five years. In this herd of 40 animals several were below the line of profit the first year. After the second year as a result of culling, all animals were kept at a good profit.

TABLE 4

COMPARISON OF THE FIVE MOST PROFITABLE AND THE FIVE LEAST PROFITABLE COWS IN THE STORRS EXPERIMENT STATION HERD FOR FIVE YEARS [1]

YEAR	COST OF FOOD	YIELD OF FAT IN LB.	PROFIT
1899			
Five most profitable cows	$56.54	304.2	$26.91
Five least profitable cows	52.02	188.6	−4.09
Difference	$4.52	115.6	$31.00
1900			
Five most profitable cows	$60.30	377.4	$43.27
Five least profitable cows	45.38	164.4	−5.75
Difference	$14.92	213.0	$49.02
1901			
Five most profitable cows	$53.24	375.3	$44.25
Five least profitable cows	43.38	217.2	15.68
Difference	$9.86	158.1	$28.57
1902			
Five most profitable cows	$59.52	376.2	$43.71
Five least profitable cows	51.45	236.6	13.71
Difference	$8.07	139.6	$30.00
1903			
Five most profitable cows	$59.46	365.5	$40.23
Five least profitable cows	56.11	268.9	17.67
Difference	$3.35	96.6	$22.56

Here we find the average cost of feeding the most profitable cows during five years was $8.14 each per year more than the cost of feeding the five least profitable cows for the same length of time, while the difference in the production of the

[1] Bulletin No. 29.

two groups averaged 144.5 pounds of fat. In this case the additional fat produced by the best cows cost only 5.6 cents a pound.

The best Jersey cow in the University of Missouri herd averaged 9,289 pounds milk and 437 pounds of butter fat for five years, as compared with an average of 2,503 pounds of milk and 122 pounds of butter fat for 3 years by the most inferior. The best Holstein averaged 15,525 pounds of milk for five years, in contrast to 5,681 pounds for three years by the poorest.

The data below are from the records of the Iowa Experiment Station.[1]

TABLE 5
From Four Years' Records of the Iowa Experiment Station

Description of Cow	Lb. of Milk Produced per Year	Per cent Fat	Lb. Butter Fat Produced per Year	Cost of Feed
Best Holstein	12,111	3.81	461	$29.83
Poorest Holstein	6,667	3.16	211	21.71
Difference	5,444		250	$8.12
Best Shorthorn	9,869	4.12	406	$27.38
Poorest Shorthorn	3,059	3.50	107	23.83
Difference	6,810		299	$3.55
Best Red Poll	7,225	4.29	310	$25.32
Poorest Red Poll	5,249	3.85	202	25.24
Difference	1,976		108	$00.08
Best Jersey	6,523	7.00	456	$26.26
Poorest Jersey	4,087	4.94	202	18.54
Difference	2,436		254	$7.72

[1] Data prepared by the writer but not published.

The data given include four years, 1897–1900, and the feeds used were charged at the current market price, which was very low at that time. The difference in cost of feed between the best and poorest Holstein was $8.12, while the best cow produced 250 pounds more fat. The difference in cost of feed for the best and poorest Shorthorn was $3.55, and one exceeded the other 299 pounds in butter fat. In the case of the Jerseys the difference in the cost of feed was $7.72, and in the amount of fat produced 254 pounds. In the data given from the Iowa Experiment Station it cost from $5 to $12 more to feed the cow that produced 300 to 400 pounds of fat than it did to feed the ones producing 150 to 250 pounds.

At the Storrs Experiment Station the most profitable cows were fed $8.14 worth of feed in excess of that given the least profitable. The data given from the Southern States indicate that the thirty best cows consumed $27.47 more feed and produced 291 pounds more butter fat each than did the poorest thirty animals. The general conclusion can be drawn, based upon the above, that it costs from $5 to $30 more per year to feed a cow, depending upon the price of feeds, that will produce 300 to 400 pounds of fat, than it does to feed a cow producing from 100 to 200 pounds of fat.

These figures illustrate very closely the conditions as they exist in the average herd of cows that are kept for milking purposes. The farmer too often makes the mistake of thinking that a certain number of cows should give a certain amount of product, and that the sum received from his herd should be comparable with that received from another herd of the same size. The point is overlooked that it is not quantity, either of milk produced or the number of animals kept, that determines the profit, but the income in excess of the cost of production. In any herd of 15 or 20 cows that has not been carefully culled a greater total profit will be realized by retaining from one half to two thirds of the herd and disposing of the inferior ones. This is especially true of herds composed of common grade and dual-purpose animals.

One cow producing 300 pounds of fat in a year makes vastly more net profit than do two producing 150 pounds each, although the total production is the same in each case. The reason can be easily understood when we consider the use which the animal makes of its feed. The first use to which the cow puts the feed given her, as is the case with any animal, is to maintain body, which in this case is keeping the machinery for milk production. The feed necessary for this purpose is called the ration of maintenance. With an ordinary cow producing 200 pounds of fat in a year, the ration of maintenance represents from 50 to 60 per cent of the feed she consumes; in the case of a better producing cow about from 40 to 50 per cent of her ration; while an extraordinary producing cow may not use over 35 per cent of her ration for this purpose. This may be looked upon as a fixed charge which has to be made in case of every animal, and since in any, except especially good producing herds, at least half of the feed will be used for this purpose it is clear that keeping two cows to do the work of one is increasing the total amount of feed used by something like 25 per cent.

Cause of the Variation in Individual Animals. — An extensive investigation is reported by the Missouri Agricultural Experiment Station on this point.[1] Two registered Jersey cows were used; the better, known as No. 27, during the first two lactation periods produced 3.9 pounds of fat and 2.8

---

[1] Research Bulletin No. 2.

pounds of milk for each pound produced by No. 62, the inferior cow kept in the same herd. These two cows were bred so they calved one week apart, and were kept under the following conditions: —

1. Complete record kept of amount and composition of feeds consumed.

2. Ration fed the two cows of the same composition at all times, the amount varied to suit the individual.

3. Cows kept at uniform weight.

4. Complete records made of milk produced and of its composition.

5. Cows kept farrow.

6. Digestion trial conducted when the cows were at their maximum production.

7. Cows kept on maintenance for four months at the end of the milking period to determine maintenance. Maintenance ration of the same composition as that fed when producing milk to determine maintenance in terms of the ration fed.

The results were as follows: during the year, No. 27, the better cow, produced 8522 pounds of milk and 469 pounds of fat, while No. 62, the inferior cow, produced only 3188 pounds of milk and 169 pounds of fat.

(1) The maintenance requirement was practically the same for each.

(2) The coefficient of digestion was almost exactly the same.

In the following tables are given the feed consumed during the year, the amount of feed required for maintenance, and the amount available for milk production by the two cows.

SHOWING PORTION OF RATION AVAILABLE FOR MILK PRODUCTION

No. 27

IN POUNDS

	GRAIN	HAY	SILAGE	GREEN FEED
Consumed during year in milk	3424.0	2904.0	8778.0	4325.0
Maintenance for year	1200.8	1204.5	4818.0	
Available for milk production	2223.2	1699.5	3960.0	4325.0

SHOWING PORTION OF RATION AVAILABLE FOR MILK PRODUCTION

No. 62

IN POUNDS

	GRAIN	HAY	SILAGE	GREEN FEED
Consumed during year in milk	1907.0	1698.0	5088.0	2102.0
Maintenance for year	1065.8	1065.8	4292.4	
Available for milk production	841.2	632.2	795.6	2102.0

Below is given the ratio between the fat produced by the two cows and the ratio of available feed.

	No. 62	No. 27
Available feed	1	2.64
Fat production	1	2.77

The interpretation of the above is that after the maintenance is supplied practically the same amount of food was required by each for a pound of fat. The real difference between the cows was that with No. 62 the capacity for using food for producing milk was limited to about one third that of No. 27. That is to say, No. 27 was a more efficient milk-producing machine on account of a greater capacity to use food above maintenance. The food she consumed was not used to any better advantage than that eaten by the inferior cow.

No. 62 required 55.8 per cent of her ration for maintenance, and No. 27 only 35 per cent.

The problem may be illustrated graphically by the lines below: —

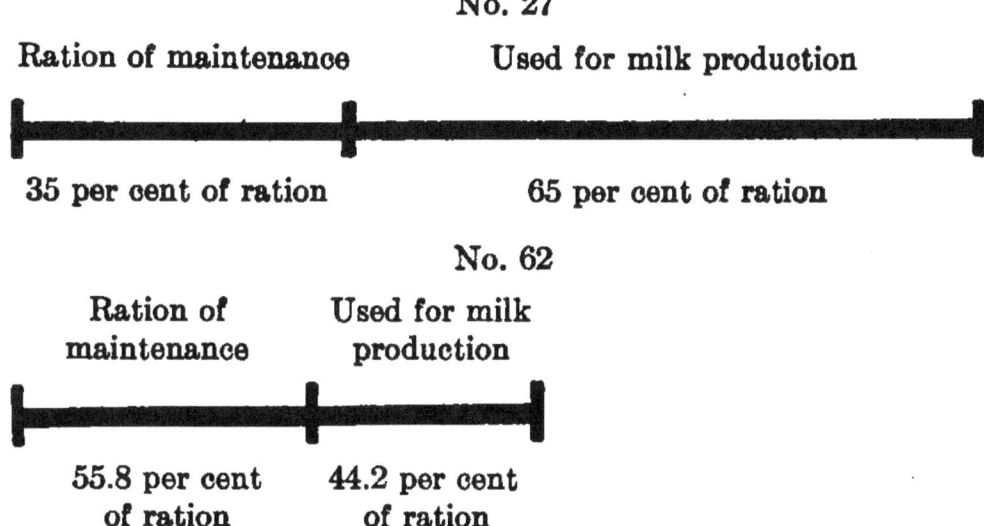

From the standpoint of economical production No. 27 is a very profitable cow, while No. 62 does not pay for her feed. In the course of the above investigations the data were compared from two other Jersey cows, and the statements given found to apply to all.

By the foregoing statement it is not meant that the superior cow had a greater digestive capacity necessarily than the inferior. The experiment above described explains how the two animals used the feed given them, but it does not indicate the real reason why the one animal secreted three times as much milk as the other. Physiologists have made many researches with the object of determining what governs the secretion of milk. The matter is far from being understood at present. It is probable, however, that a substance is secreted in the body somewhere, possibly by some part of the reproductive organs, that circulates with the blood. This substance, whatever it may be, stimulates the udder gland to produce milk. The amount of this substance that a certain animal will secrete is hereditary, although influenced to some extent by the feed and other conditions. A good dairy cow is one that has inherited the power to form a large amount of this substance. An inferior dairy cow, like No. 62 in the experiment reported, is one that does not produce a sufficient amount of this stimulating principle. From this viewpoint the selection of the cow is choosing the one that has a strong stimulation to produce milk.

Another experiment made by the writer illustrates how strong this stimulation to produce milk is with a highly developed dairy cow. A mature Jersey cow was fed a liberal ration while dry; she calved in more than moderate flesh. Her ration was then adjusted to supply only nutrients sufficient to support the body, leaving nothing for producing milk. She was compelled either to cease producing milk or to make it from the reserve on her body. This was continued for 30 days. At the end of this time she was producing only one pound of milk per day less than in the beginning, but was so weak she could hardly get up without assistance. During this time she lost 115 pounds in her weight, and produced over 90 pounds of solids in the milk from her own body.

According to the view expressed the high-producing cow secretes milk on account of the strong stimulation she has in her body. To replace these nutrients she has a keen appetite, and consumes a large amount of food. The point is, she consumes the heavy ration because she produces a large amount of milk. The consumption of the heavy ration is the result and not the cause of the heavy milk production. This subject is further discussed in connection with feeding, to which it bears an important relation.

CHAPTER XII

HOW INDIVIDUAL SELECTION IS MADE

There are two means of selecting dairy cows to obtain the characteristics desired: that is, individuals capable of using the maximum amount of feed for milk production. The first of these is by the conformation or type of the animal, and the second is by keeping records of the production of each individual animal.

There is no doubt that there is a dairy form or type that generally goes with large milk and butter production. This type is sufficient, as a rule, to enable competent judges to select very good cows from inferior ones, as described previously. However, dairy type alone cannot be depended upon as a means of selecting the best dairy animals from among a number of good ones. As a rule, it is possible to select cows that are capable of producing from 300 to 350 pounds fat in a year from those that will produce one half that amount, but it is practically impossible to select a cow capable of producing 400 to 500 pounds of fat from one producing 300 pounds of fat. Often even those most familiar with the subject will make decided errors in selecting animals by this means. Within the limits given selection by conformation can be made use of by an intelligent and observing person.

In case an animal is to be purchased for which no records of production have been kept, the buyer must depend mostly upon the evidences of dairy characteristics as shown by the animal, and it will be well to depend upon these rather than to attempt to select by weighing and testing the milk for a single milking, or even an entire day. As a rule, in such a case the conformation of the animals should be depended upon as an indication of dairy quality rather than statements regarding the production of the animal, unless such statements are based upon actual weights and tests taken.

Selection of Cows by Test. — The only satisfactory way to select the profitable from the unprofitable in a herd of dairy cows is by keeping records of the amount of milk produced and testing for butter fat at regular intervals. While it is only occasionally that cows may be purchased at the present time that have records of milk production, there is no excuse when the animals are once in the herd for not keeping records in order that the unprofitable ones may be rejected from the herd as soon as possible. In making such records it will depend upon the use made of the milk as to what records are made. If milk is sold by measure or by weight regardless of the fat content, the producer is interested in the amount of milk produced; while if butter fat is sold, or if the price of the milk is based upon the butter fat, then the quantity of milk and the per cent of fat both need to be considered.

Overrating the Importance of Rich Milk. — A common mistake made in judging of the value of cows is attaching undue importance to the richness of the milk. The cow that gives the richest milk does not necessarily produce the largest quantity and is not necessarily any better or even as profitable as the cow yielding milk with a smaller per cent of fat. It is the total fat production that counts. The fallacy of depending upon the per cent of fat in the milk as an index of the value of the cow is shown clearly by comparing the records of the best Holstein cow and the best Jersey cow as given in Table 5 from the records of the Iowa Experiment Station. There will be noticed that in this case the per cent of fat in the milk of the best Jersey was 7 per cent, while that of the Holstein was 3.81 per cent. Judging by the per cent of fat alone, the Jersey would have been counted a much better animal, while the total fat production for the year was 5 pounds less with the 7 per cent of fat than with the comparatively low per cent of fat of the Holstein. In this case, if the animals had been judged by the amount of milk produced, the best Holstein would have been counted much better than the best Jersey, as she produced nearly double the amount of milk.

Table 6 is compiled from the records of the University of Missouri herd. The milk, per cent of fat, and total fat production for a year is given for some of the best and some of the poorest in each breed.

These figures show that high fat production is not necessarily connected with a high average per cent of fat. In fact, as a rule the highest production of fat is accompanied with a per cent of fat average for the breed or lower. No general rule can be drawn on this point. In many cases the cows having the lowest fat production in the herd have the highest fat percentage, while in other cases the fat content of their milk is also low. The point to be kept in mind is that cows vary far more in total milk production than they do in the fat content.

In the figures given, one Jersey produced over five times as much milk as another, and this is not unusual with any breed. It might be possible to find one cow that would show an average fat content double that of another in the same breed, but such extremes would be very exceptional. In a pure-bred herd the average per cent of fat seldom varies more than one fourth from the lowest to the highest. The fat content of the milk of each individual cow becomes of more importance when herds of mixed breeds are used than is the case where the blood of one breed predominates in all. This comes about on account of the wide variation in fat content between breeds; and in animals of mixed breeding it is impossible to tell without testing what characteristic has been inherited regarding the fat content of the milk.

TABLE 6

RELATION OF PER CENT OF FAT TO TOTAL FAT PRODUCTION

UNIVERSITY OF MISSOURI HERD

BREED	HERD NO.	YIELD MILK PER YEAR	TOTAL FAT YIELD	AVERAGE PER CENT FAT
Jersey	1	13,895	681	4.90
Jersey	16	12,729	634	4.98
Jersey	124	13,322	625	4.69
Jersey	62	3,188	169	5.31
Jersey	5	2,797	176	6.20
Jersey	45	2,849	126	4.60
Holstein	204	18,405	618	3.41
Holstein	211	17,692	519	2.93
Holstein	214	5,436	212	3.92
Holstein	205	6,387	208	3.20
Shorthorn	403	13,789	490	3.55
Shorthorn	401	12,341	515	4.17
Shorthorn	400	4,543	195	4.20

In communities where milk is sold to factories on the butter-fat basis certain herds are often considered the best in the community simply because the per cent of fat from these herds is the highest; but this should not be taken as evidence that these are the most profitable herds.

There are three things that should be known in order that the relative profits may be known from the individual animals. These are the amount of milk produced, the per cent of fat, and the amount of feed consumed. In regard to variations between individual animals, these three factors stand in the order named.

Keeping Complete Milk Records. — The only plan of keeping milk records which is entirely satisfactory is that of keeping complete daily records of each individual animal. This appears to be a very large undertaking to the dairyman who has never followed such a plan, but it does not require as much work as is usually anticipated, and the advantages which follow are sufficient to justify the expense of labor required. A pair of spring balances should be provided and hung at a point convenient for the milkers and a suitable milk sheet placed on the wall beside the scales. One of the advantages of keeping complete daily records of milk production is that it makes possible the feeding of the individual cows with the greatest economy. It is impossible to feed economically unless the amount

Fig. 31. — Scales for weighing milk.

of milk produced by the individual is taken into account; and unless a daily record is at hand, there is no basis upon which to estimate the amount of feed each individual requires. Daily records also enable the herdsman to detect sickness quicker than would otherwise be the case. If there is a noticeable decline in the amount of milk produced, with no apparent cause, it is certain the animal is not in the right condition and will probably show a more marked case of sickness very soon, if not properly treated. When such a sudden decline occurs, the herdsman, by adjusting the ration and giving the cow some special attention, will be able in many cases to prevent the development of what might be a serious case of sickness.

Again daily weighing makes it possible to judge of the work done by different milkers. It is a well-known fact that some milkers are able to secure much more milk from the same cows than are others. This difference may be as much as 25 per cent, or even more. Especially where there are several milkers in the same herd, it is impossible to form a fair estimate of their work unless each man milks the same animals regularly and each lot of milk is weighed and recorded.

Without records certain cows often become favorites of the milkers for some reason, either on account of the disposition of the animal or the easy milking; and these favorites are held to be the best cows by the milker; and often the truth regarding their value cannot be told except from the records. It is the experience of all who have adopted the plan that the advantages above enumerated more than pay for the extra time required. The greatest advantage, of course, is making it possible to know the profitable cow and dispose of the inferior animals.

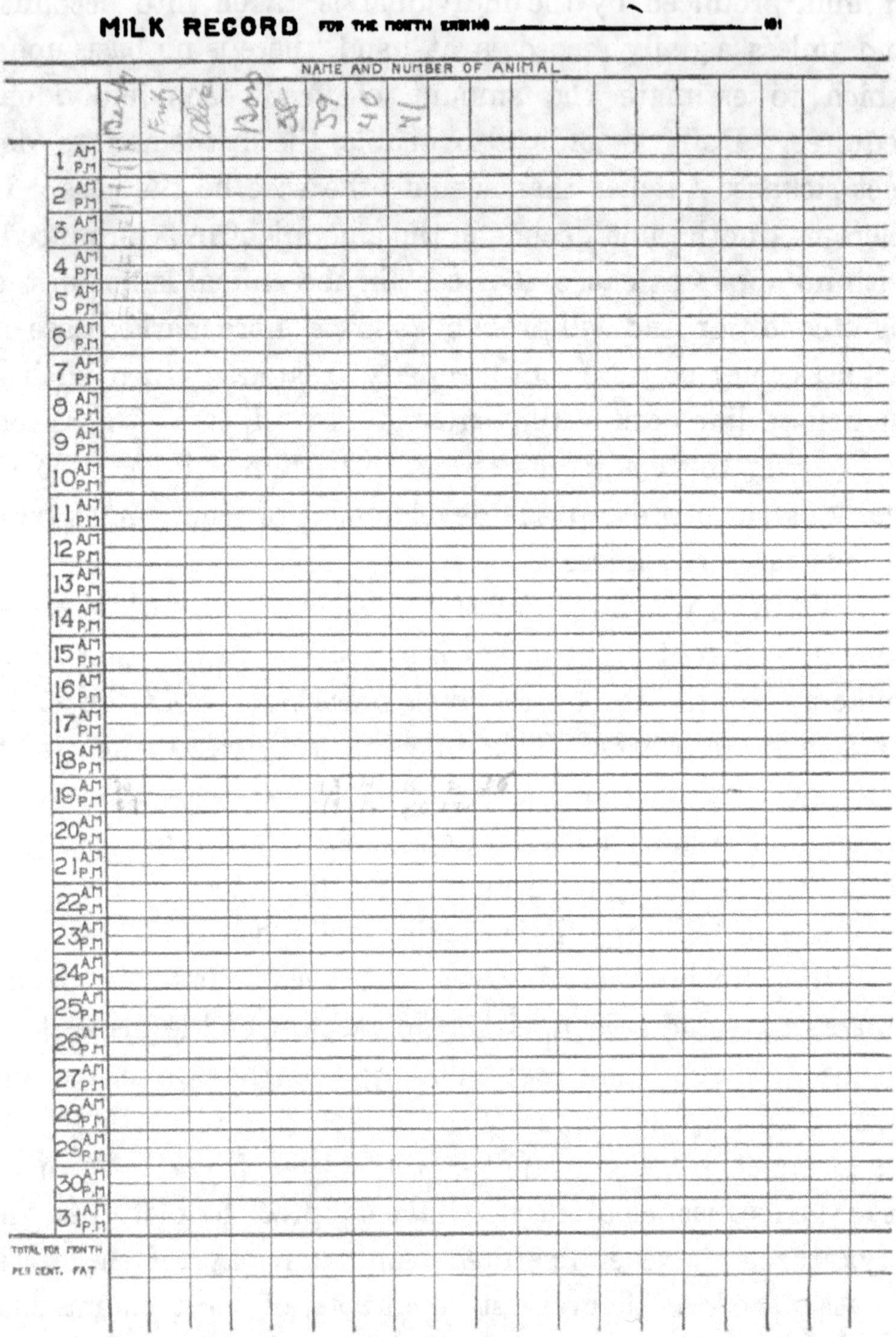

138 DAIRY CATTLE AND MILK PRODUCTION

Fig. 32. — Milk sheet. A form for keeping daily records of 18 cows for one month.

When complete milk records are kept, as is recommended, it is sufficient to take a sample covering from three to five days each month for the butter-fat test. The best plan is to take a mixed or composite sample, as it is called, for three days in the middle of the month. This is tested for butter fat, and the result counted as the average for the month.

Monthly Tests. — Another plan of keeping records of individual cows, which is well adapted for use by those who feel they cannot spend the time necessary to take daily milk

RECORD FOR MONTH OF _____

	Cow No. 1	Cow No. 2	Cow No. 3	Cow No. 4	Cow No. 5	Cow No. 6	Cow No. 7	
14 AM								
14 PM								
15 AM								
15 PM								
16 AM								
16 PM								
TOTAL								
PER DAY								
TOTAL MONTH								
% FAT								
TOTAL FAT								

Fig. 33. — Form for keeping milk records, weighing three days per month.

records, is to weigh the milk and make tests for per cent fat at monthly intervals. In carrying out this method, the best arrangement is to weigh the milk from each individual cow for three days about the middle of the month. A composite or average sample of the milk is also taken during the same time, which is tested for per cent of fat. The average milk and fat produced for the three days is taken as the average for that month. This should be carried out regularly each month during the year. The total production of each cow as shown by such tests is close enough to the actual for all practical purposes. The objections to this plan are that the owner often forgets to take samples when it is not done regularly. The advantages of having daily milk records as guides for feeding and for checking up milkers should be taken into account.

Fig. 33 shows a suitable form for keeping the weights for the three days. The totals for the month can be kept in the form illustrated in Fig. 34.

Testing at Intervals of Three Months. — Another method which has less value than either of the foregoing, but is better than no testing at all, is to make the weights of the milk produced by each animal and a sample covering seven days' time at intervals of three months, then from this week's average calculate the production for three months, of which this is the middle week.

Method of Taking and Testing Samples. — Where many samples are to be taken, the most convenient and accurate method is with a sampling tube which can be purchased from a dairy supply company. In taking a sample with this arrangement, the sampler, which is a brass or copper tube, is lowered in the pail of milk and then closed. The milk remaining in the tube is placed in a pint glass jar bearing the name or number of the cow. These jars must be kept covered tightly to prevent evaporation. If a sampling tube is not at hand, a fairly accurate sample may be prepared by taking equal quantities of the milk from each milking with a very small dipper and placing in a jar provided for that purpose. Some preservative is used to prevent the souring of the milk before the time for testing. For this purpose formaline, which may be purchased from any drug store, is one of the best agents. Ten drops of this substance will be sufficient to preserve half a pint of milk for several days. A small amount of corrosive sublimate serves the same purpose. This substance is a strong poison, and for this reason milk containing it must be handled very carefully. A quantity as large as a grain of wheat is ample to preserve a sample for testing. It is well to use prepared tablets containing a coloring matter, or to add some common dye to prevent the milk from being used for food by mistake.

The samples when complete are to be tested by the Babcock test. The reading of the test is per cent butter fat, or pounds of pure fat per 100 pounds of milk. The total yield of milk multiplied by the per cent of fat gives the total fat yield. If the cream from this milk should be separated and churned into butter, a larger quantity of butter would be obtained than there was of butter fat. This results from the incorporation of some curd, considerable water, and a small amount of salt with the fat, the mixture being ordinary butter. If proper methods are followed, the yield of butter will exceed that of fat by about one sixth, which is known as the overrun. In estimating the amount of butter from the butter fat it is customary to add one sixth to the butter fat.

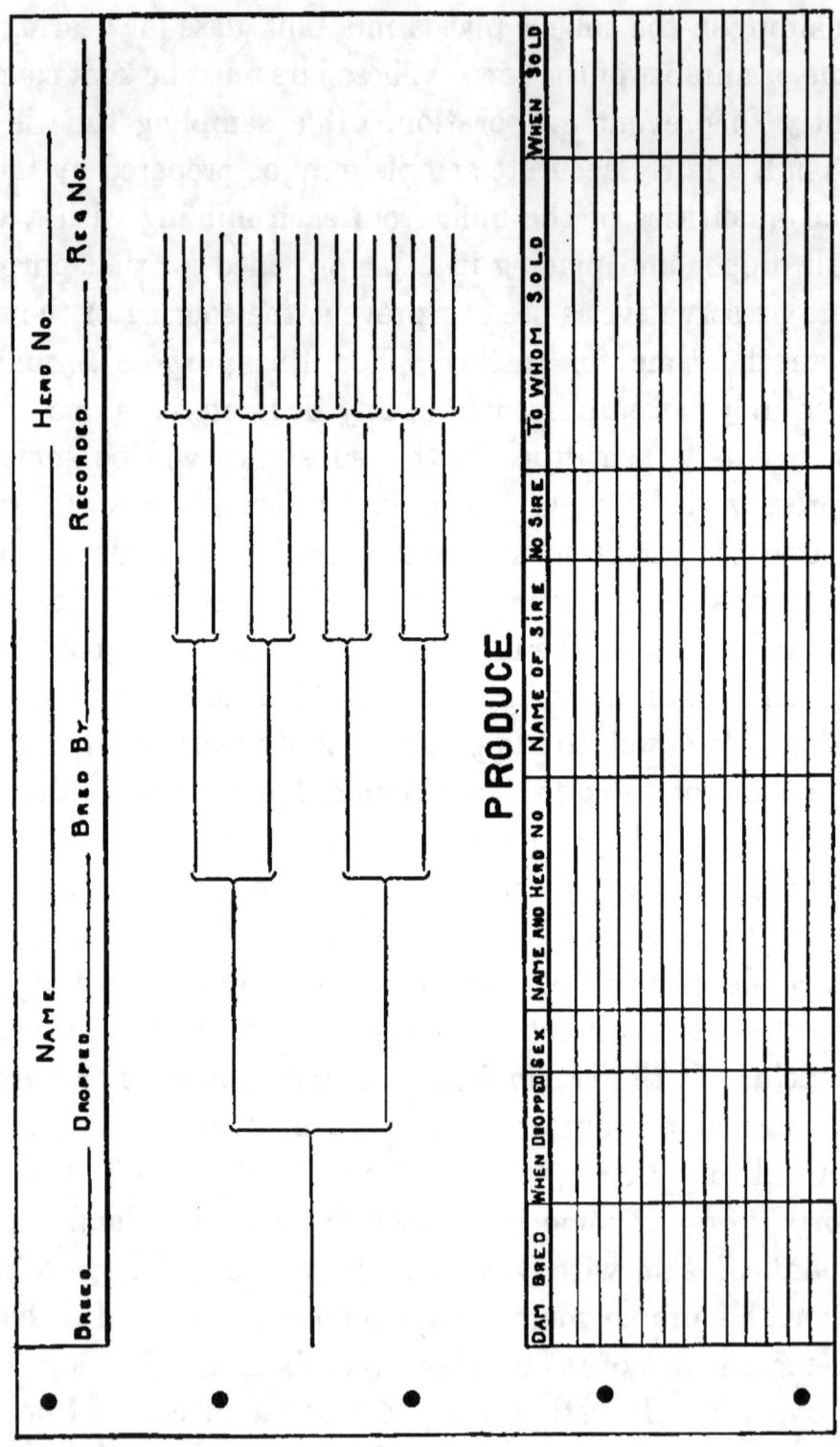

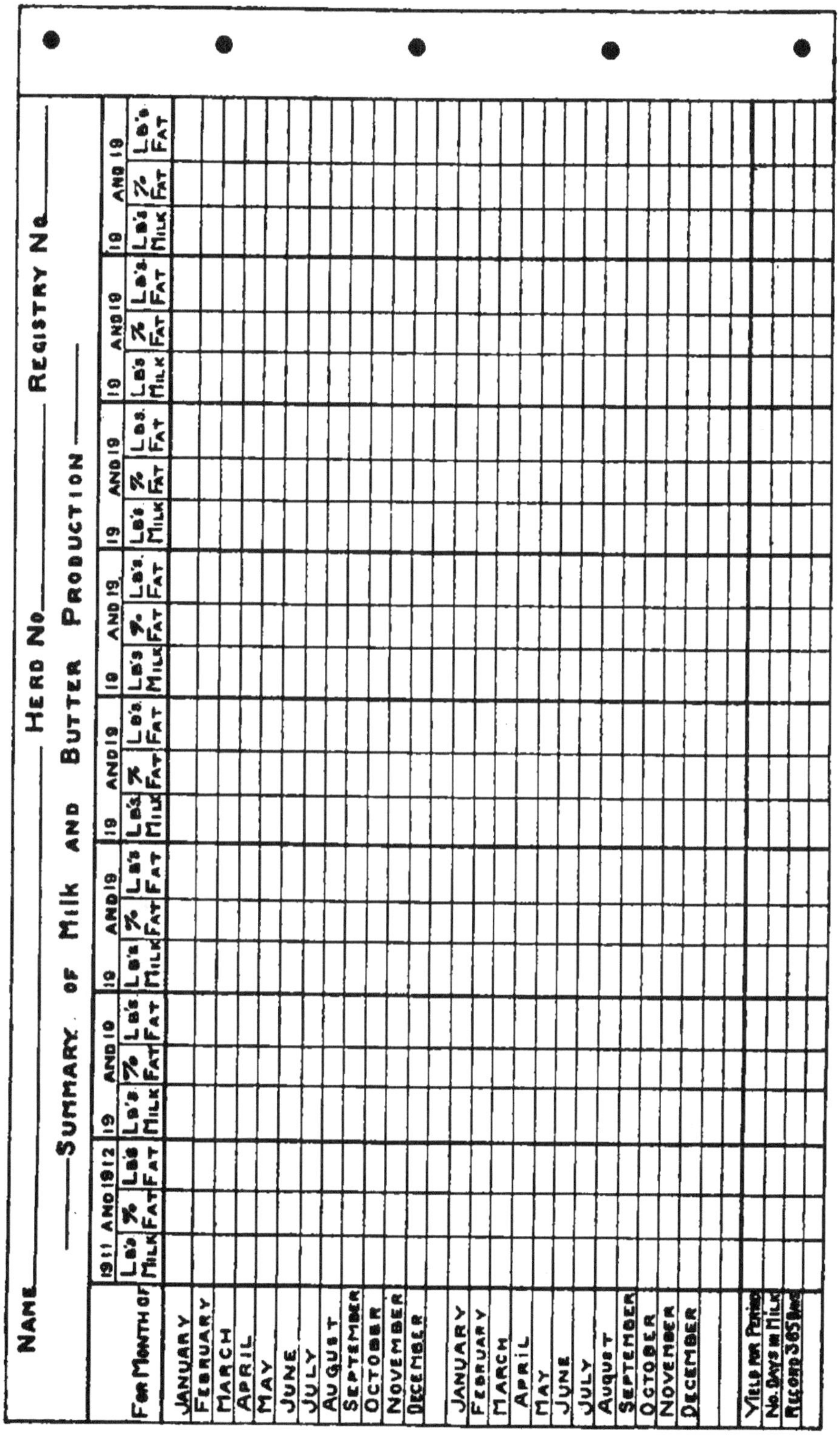

NAME ______ HERD No. ______ REGISTRY No. ______

SUMMARY OF MILK AND BUTTER PRODUCTION

For Month of	1911 and 1912			19 and 19			19 and 19			19 and 19			19 and 19			19 and 19		
	Lb's Milk	% Fat	Lb's Fat	Lb's Milk	% Fat	Lb's Fat	Lb's Milk	% Fat	Lb's Fat	Lb's Milk	% Fat	Lb's Fat	Lb's Milk	% Fat	Lb's Fat	Lb's Milk	% Fat	Lb's Fat
January																		
February																		
March																		
April																		
May																		
June																		
July																		
August																		
September																		
October																		
November																		
December																		
January																		
February																		
March																		
April																		
May																		
June																		
July																		
August																		
September																		
October																		
November																		
December																		
Yield for Period																		
No. Days in Milk																		
Record 365 Days																		

FIG. 34. — Form for a permanent record.

In examining records of dairy cows, care should be observed to distinguish between butter fat and butter records.

Form of Keeping Record. — In keeping records of the production of individual animals one of the first essentials is to arrange a system to be followed. In many cases those who have undertaken to keep records of milk and fat production have not carried out their intentions long, for the reason that they started without any carefully prepared plan and with no convenient way to keep their records. Figure 32 shows a convenient form of arrangement for recording the daily milk yield for a month. As a rule, only the totals by months are used for references later. A permanent book should be provided for preserving these totals in a convenient form. Figure 34 illustrates a satisfactory plan used by the University of Missouri since 1892.

This form of record is especially valuable when the herd consists of pure-bred animals. A book is ruled or purchased, having the two pages, as illustrated in Fig. 34, opposite each other. On the left page are recorded the pedigree of the animal and the record of calves. On the right-hand page are recorded the yield of milk by months, the per cent of fat, and the total fat yield. It will be noted that the months of the year are arranged in order, 24 months being found in each column. In entering a year's record of a cow, the first month's record is entered opposite that month the first time it appears, beginning at the top of the column. It leaves room in every case for at least 13 months' record.

At the end of the milking period the record is added and the total inserted. In finding the average per cent of fat, the total fat yield should be divided by the total milk to give the true average per cent of fat. An average found by add-

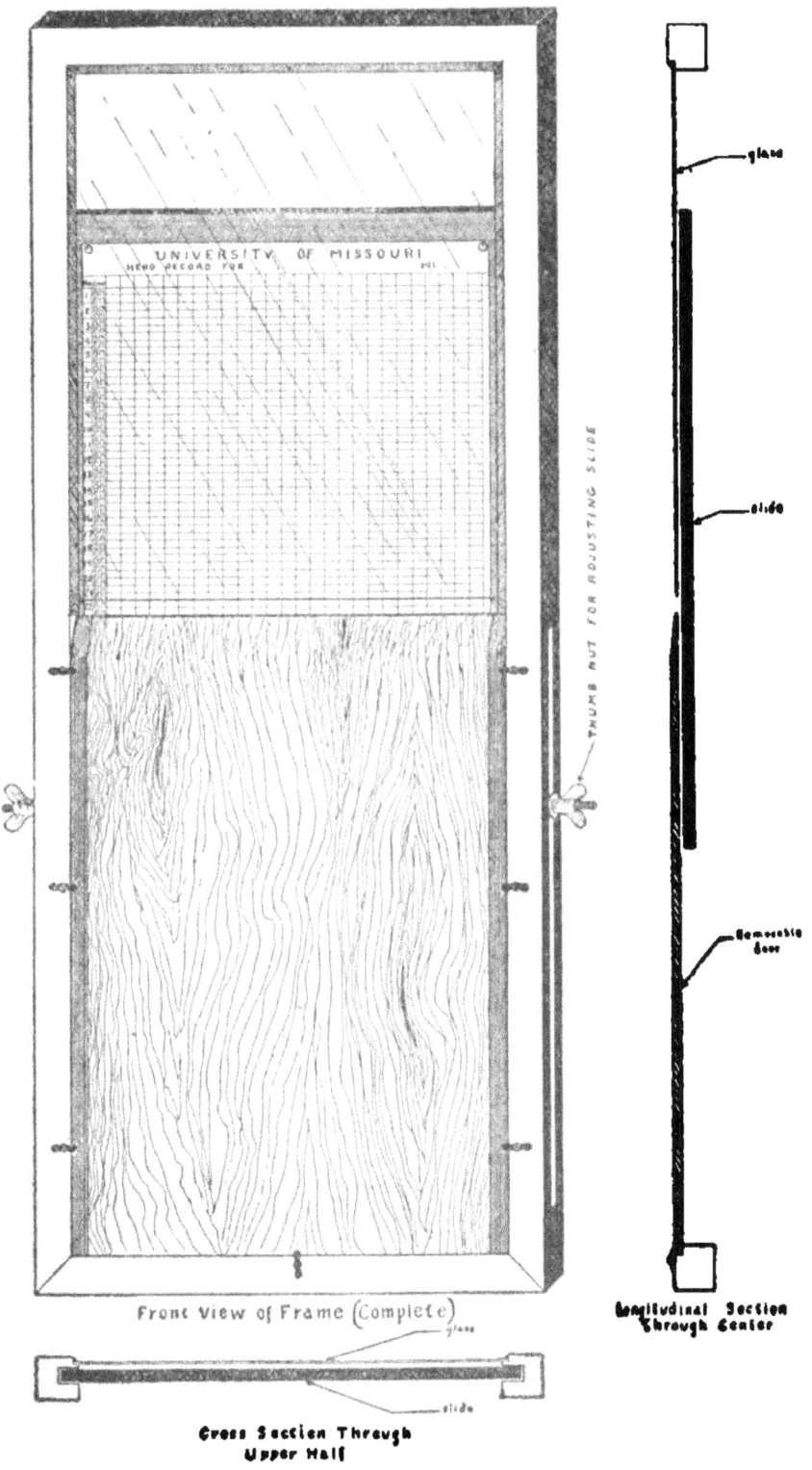

Fig. 35. — Frame for holding milk record sheets.

ing the monthly tests, and dividing by the number made, does not give a true average, but usually a figure somewhat too high, since the amount of milk usually is small when the per cent of fat runs the highest.

Danish Test Associations. — So far the American farmer has been slow to carry out the details of keeping records of individual cows in his herds, even when he is convinced that it should be done. In Denmark, where the same necessity for selecting individual cows was recognized earlier and was even more important than here on account of the higher priced land and more expensive feeds, a system of coöperative test associations has been adopted, which seems to solve the difficult problem of getting this work done. The first cow-testing association was established at Askov, Denmark, in 1895.[1] The plan was at once successful, and spread rapidly over Denmark and into other nations as well. The following data[2] give the number organized up to 1904.

	Denmark	Sweden	Norway	Germany	Finland
1895	2				
1896	13				
1897	15			1	
1898	58	1	1		1
1899	82	7	5	1	1
1900	49	20	13	1	
1901	41	42	46	5	1
1902	47	66	42	9	3
1903	55	51	30	11	5
1904	28	86	23	34	10
Total	390	273	160	62	21

[1] Buer, Die danischen Kontrollvereine.
[2] Bul. 137, Michigan Dairy and Food Department, 1907.

The organizations are coöperative in nature, but receive a subsidy of about $60 per year from the government. The organization is formed by the farmers of a neighborhood owning from 300 to 1000 cows. A man is employed, usually a student from an agricultural school, to do the testing. The tester, or Control Assistant, as he is called, is also expected to be able to advise the farmers regarding the feeding and breeding of the dairy cattle. Each herd is visited in turn, the frequency depending upon the number; the milk is weighed for one day, the fat content determined by the Gerber method, and a record of the feed consumed by each animal is taken. All records are kept for the owner by the tester.

The Danish system of expressing cost of production in feed units is followed, and the records show the amount of milk and fat produced by each cow, the amount of feed required, and the results per 100 feed units.

The following data illustrate the remarkable results that have been obtained in certain herds.

HERD B., OWNED BY AUG. KINCK, BELTABERGA, SWEDEN [1]

Testing Period 365 Days	Av. No. of Cows	Av. Feed Units per Cow	Av. No. Lb. Milk per Cow	Av. Test	Av. Lb. Fat per Cow	100 Feed Units Gave	
						Lb. Milk	Lb. Butter
1899–1900 .	70	2421	7,320	3.05	223	302	10.1
1900–1901 .	28	2695	7,905	3.13	247	272	10.1
1901–1902 .	46	2566	9,003	3.20	288	350	12.3
1902–1903 .	55	2507	9,984	3.18	318	398	13.9
1903–1904 .	61	2587	10,584	3.22	341	407	14.5
1904–1905 .	64	2743	11,236	3.22	362	409	14.5
1905–1906 .	71	3035	11,333	3.21	364	372	13.2
Increase .		614	4,013		141	70	3.1

[1] Data from H. Rabild, Dairy Division, U. S. Dept. Agric.

Such remarkable results are made possible by the farmer knowing exactly what profit each animal has made; and naturally the unprofitable ones are disposed of as soon as possible. As a result, the average production of milk for the entire country of Denmark has increased surprisingly within a few years. It is safe to say the improvement in production has been faster in Denmark during the last 15 years than in any other part of the world. The cost of the Test Association per cow in Denmark varies from 30 to 45 cents in addition to the board of the tester. The tester, or Control Assistant, receives from $100 to $150 per year in addition to his board, which is supplied by the farmer for whom he is testing. The positions are much sought after, on account of the excellent opportunity to gain good practical experience.

The first coöperative Cow Test Association in America was established at Fremont, Newaygo Co., Mich., in 1905. Up to the present time a number of such associations have been formed and are meeting with a fair amount of success. The plan followed is patterned after the Danish, but the cost per cow is higher on account of the higher scale of wages. The total cost need not exceed $1 to $1.50 per cow annually in addition to the board of the tester.

INFLUENCE OF THE AGE OF THE COW ON THE YIELD AND RICHNESS OF MILK

One question of considerable importance that arises in connection with the selection of the individual cow is the influence of the age of the animal on the richness and yield of milk. The dairyman finds a certain heifer, for example, to have an unusually high or low per cent of fat or yield of milk, and the question at once arises as to what she can be expected to do when mature. The following data and discussion are based upon a study made by the author of the complete records of a Jersey herd covering 15 years, and of a Holstein herd for 10 years. To this is added a table compiled from the published official records of the four leading dairy breeds.

It must be understood first of all that the figures given are averages, and that some individual animals will vary widely from the average. The data will, however, give a fairly good index of what may be expected, as it may be reasonably assumed that the majority will follow near the average.

Table 7 is compiled from the records of the herd owned by the University of Missouri. These figures are nearly all from Jersey cows. It has been the custom in this herd

TABLE 7

EFFECT OF AGE ON THE QUANTITY AND QUALITY OF MILK.
UNIVERSITY OF MISSOURI HERD

Lactation Period No.	41 Cows HAVING 3 PERIODS		28 Cows HAVING 4 PERIODS		21 Cows HAVING 5 PERIODS		16 Cows HAVING 6 PERIODS		10 Cows HAVING 7 PERIODS	
	Av. Lb. Milk	Av. Per Ct. Fat	Av. Lb. Milk	Av. Per Ct. Fat	Av. Lb. Milk	Av. Per Ct. Fat.	Av. Lb. Milk	Av. Per Ct. Fat	Av. Lb. Milk	Av. Per Ct. Fat.
1	4912	4.87	4957	4.90	4755	4.95	4711	4.90	4754	4.85
2	5716	4.87	5597	4.98	5477	5.00	5239	5.05	5060	5.10
3	6601	4.90	6139	4.95	5887	5.00	5707	5.00	5648	5.09
4			6262	4.81	6318	4.80	6266	4.77	6575	4.82
5					6159	4.88	6169	4.76	6341	4.68
6							6511	4.60	6339	4.59
7									6282	4.57

to have heifers drop their first calf when from 24 to 27 months of age, so "Lactation Period No. 1" as given in the table is the two-year-old record.

The "lactation period" as used here is the entire milking period, if it be less than 12 months; if the milking period extended beyond a year, the figures used are for the first 12 months only. The amount of milk is the total from weights taken of each milking separately. The per cent of fat was taken each month from a composite sample covering ten milkings in the middle of the month.

In Table 7 the records are first given of the yield of milk and per cent of fat by lactation periods for 41 cows, of which records are at hand for the first three milking periods. Then follow the records of 28 cows for the first four periods, 21 cows for five periods, etc.

Table 8 gives the data for the Jersey herd, regarding the fat yield in another form. Here the highest yield of fat is

TABLE 8

JERSEY HERD

UNIVERSITY OF MISSOURI

Fat Production by Years expressed in Per Cent

Number Milk Period	Number Cows	Fat Yield in Per Cent
1	37	71
2	37	84
3	37	93
4	26	100
5	20	98
6	15	98
7	10	95

expressed by 100, and the yield for other periods of lactation expressed by per cents. For example, in Period 1, at the age of two years, 37 cows averaged 71 pounds of fat for each 100 they produced in the fourth milking period; and in the seventh period 10 cows averaged 98 pounds of fat for each 100 the same cows produced in the fourth period.

Table 9 is compiled from the Records of Official Testing by the four leading dairy breeds. The Holstein figures represent the seven-day tests recorded for the year ending June, 1909. The figures for the other breeds include all yearly records available to the writer at the time the compilation was made.

This table gives the number of cows of each age included, the fat record expressed in per cent, counting that produced at five years or above as 100, and the requirements for entry to the various Advanced Registries at different ages expressed in per cents, with the requirement for full age as 100.

TABLE 9

OFFICIAL TESTS

Yield Fat expressed in Per Cents, Comparison with Rules

Age Yrs.	Holstein			Jersey			Ayrshire			Guernsey		
	No. Cows	Fat Yield in Per Cent	Rules	No. Cows	Fat Yield in Per Cent	Rules	No. Cows	Fat Yield in Per Cent	Rules	No. Cows	Fat Yield in Per Cent	Rules
2	630	65	60	55	68	65	39	70	66	232	77	70
3	246	82	75	43	78	75	16	85	73	88	83	80
4	200	91	86	16	84	87	10	97	87	70	91	90
5	684	100	100	97	100	100	51	100	100	311	100	100

The Yield of Milk. — It will be seen from Table 7 that in the University of Missouri herd the highest yield of milk was in the fourth milking period, which corresponds closest to the sixth year. The two-year-old heifers on the average yielded 71 pounds of fat for each 100 when mature; the three-year-olds 84 pounds, and the four-year-olds 93 pounds.

From Table 9 we see that with the Ayrshires the two-year-old cows produced 70 per cent, the Holsteins 65 per cent, the Jerseys 68 per cent, and the Guernseys 77 per cent of the average production for the mature cows. In each case it might well be observed that the two-year-old animals produced more in proportion to the mature cows than required by the rules of the breed association under which the tests were made. The chances are evidently better for a cow to enter any of the Advanced Registries as a two-year-old than when mature.

The decline in milk production with age is difficult to represent, since it varies widely. In the table given from the university herd up to the seventh period, the decline on the average was slight. Many of the cows whose records are used did their best year's work in the seventh period or later. Since the number having more than seven periods in our herd is limited, the figures are not given beyond that point.

Probably the majority of dairy cattle are rejected from the herd on account of failure to breed, or from udder trouble, before the effect of advancing years can be observed to any marked extent. It is a fact often observed that a cow may make her best record when 10 or 11 years old, although as a rule she does her best rather earlier. If a dairy cow continues to breed, she usually shows no marked decline until at least 12 years old. Occasionally a cow continues to breed until she is 16 or 18 years old.

The Per Cent of Fat. — Table 7 gives the average per cent of fat by lactation periods. This average is the true average found by dividing the total fat by the total milk produced. It will be noted that there is little change during the first three milking periods; but from that time on there is a slow but slight decline with advancing years. This agrees with the results found by Hills.[1] There are several conditions that may bring about considerable of a variation from the average with an individual, but these variations are naturally much less than is the case with the milk yield. While the daily variations in the per cent of fat in the milk of all cows are constant and striking, the average for the entire milking period varies but little from year to year. While it is not entirely safe to judge the future milk production of a cow from her two-year-old record, it is reasonably safe to judge the richness of her milk.

The whole subject may be summed up as follows: —

A dairy cow on the average as a two-year-old may be expected to produce about 70 per cent; as a three-year-old around 80 per cent; and as a four-year-old about 90 per cent of the milk and butter fat she will produce under the same treatment when mature.

The richness of milk remains practically constant from year to year, except that after the third milking period there is a slow, gradual decline with advancing years.

[1] 20th Annual Report, Vermont Exp. Sta.

CHAPTER XIII

SELECTION OF THE HERD BULL

CARE AND MANAGEMENT

It has long been an axiom of the breeder that the sire is half the herd, and it is generally accepted as a fit expression of an important rule. The skillful breeder of any kind of stock does not need to have it pointed out to him how important it is that the sire be properly selected. If he is a skillful breeder, it is largely because he realizes the importance of the sire and knows how to select him. While the skilled breeder realizes the importance of this in breeding, the average dairyman does not give the question of the selection of the sire one tenth the attention the importance of the question demands. Thousands of men make use of a scrub or grade sire on account of mistaken economy in cost, rather than pay a few dollars more for an animal that is almost certain to transmit desirable qualities.

First of all, the bull selected should be a pure-bred of the breed to which the cows belong, or, in case grading up is to be done, of the breed selected as best suiting the purpose. It is easily understood that a bull whose ancestors have been bred for many generations for one purpose is more certain to transmit that character than one whose ancestry is mixed. As already pointed out, the dairy cow of to-day has been developed until she produces many times the product of her ancestors, and is really abnormal in this respect. This power of producing enormous quantities of milk, being an acquired characteristic, is easily lost; and in the case of cows many revert to the older type, and are the unprofitable producers that have to be culled constantly from any herd.

Since the producing of large quantities of milk is not natural, but acquired, it is only by constant selection that this characteristic can be retained. Unless selection is made, not only will no progress be made, but the tendency will be backward. To even retain the milking qualities of a good grade herd at a uniform level, it is necessary to use a bull better bred than are the cows. This makes it imperative, in all herds where any attempt is made to advance, or even retain a high dairy quality, that the sire be selected with great care, especially regarding the dairy qualities of his nearest female ancestors.

Almost any pure-bred bull with good milking ancestry will improve a mixed herd or one of poor dairy quality; but for the highly developed herd it is a much more serious matter to select the proper bull. The breeder of high-class purebred animals recognizes the highly important fact that a bull may not transmit the desirable qualities of his ancestors to the full degree. This class of breeders are always anxious to make use of an animal known to transmit the qualities desired. An animal that transmits characteristics with uniformity to its offspring is said to be prepotent.

The remarkable variation in the transmission of dairy qualities by different bulls is shown by the results from the University of Missouri Jersey herd. In 1884 this institution purchased four registered Jersey cows, and the entire herd on hand at the time this compilation was made (1910) was descended from these cows. Herd bulls were of course purchased from other herds, but no females. Since 1892 complete milk and fat records have been kept of every cow. Up to 1901 practically every female was retained in the herd, regardless of her dairy qualities. While the conditions under which the herd was kept were not entirely uniform, still no more variation occurred than would be the case in a private herd. A comparison is made of the records of the daughters of each bull, compared with the production of their dams. In most cases the figures given represent the whole lifetime of both. In case only a limited number of records are available, these are compared with the corresponding ones of the dam. When a lactation period extends beyond a year, only the first twelve months' record is used. The low average production is accounted for by the fact stated that for the greater part of the time all females were retained, and by the records of inferior cows sold while young.

The first bull used was Missouri Rioter, a son of Bachelor of St. Lambert. There is no record indicating the dairy quality of his dam. His sire is the only animal in his pedigree known to be a strong breeder. This bull left four daughters in the herd that have a total of 26 lactation periods, from dams with 23 periods on record. Below is given a summary of the records of these daughters and their dams.

	Dams	Daughters
Average milk yield	5380	4381
Average per cent fat	4.35	4.93
Average yield of fat	234	216

The average production of the daughters was 1009 pounds of milk and 18 pounds of fat per year below that of their dams. In every case the daughter was inferior to the dam. While the number of daughters is too few to make the data conclusive, it is certain this bull did not transmit the qualities wanted. The same results in a herd having thirty of his daughters would mean over 30,000 pounds of milk and 540 pounds of fat per year less than the dams. If each daughter was milked six years, it would mean over 6000 pounds of milk less for each during her lifetime than her dam produced. In other words, it would take over six years for the daughters to produce as much as the dams would in five years.

The next bull was Hugorotus, a cheap bull without any high-class animals in his pedigree. This bull had eleven daughters, with a total of 50 lactation periods from dams with 62 lactation periods on record. The comparison below shows the results.

	Dams	Daughters
Average yield of milk . . .	4969	4576
Average per cent fat . . .	4.66	5.49
Average yield of fat . . .	231	245

The eleven daughters average 393 pounds of milk below their dams, but on account of a marked increase in the richness of the milk, gain 14 pounds in fat per year. Six out of the eleven daughters were decidedly inferior to their dams. The general results of using this bull were disastrous. In fact, the poorest animals ever in the herd were his offspring. The averages are made as good as they are only by two full sisters sired by this bull that through some "nick" proved first-class animals. When the herd was culled on milk records alone, nine out of the eleven daughters were sold. As long as this bull was at the head of the herd, the tendency was backward.

The next bull used was Lorne of Meridale. This bull had a splendid pedigree from the standpoint of records of production, and his offspring show the result. He had twelve daughters in the herd, with a total of 67 lactation periods, from dams with 66 lactation periods on record. Below is a summary: —

	Dams	Daughters
Average yield of milk . .	4559	5969
Average per cent fat . . .	4.85	4.81
Average yield fat	221	287

His daughters show the remarkable increase of 1410 pounds of milk and 66 pounds of fat per year over the dams. In only two cases out of the eleven did the daughters fall below the dams, and in one of these only slightly. In five cases the increase was over 2000 pounds per year. If thirty daughters of this bull had been milked six years, their total milk product would exceed that of their dams by over 250,000 pounds, worth $3750 at $1.50 per hundredweight. An equal number of daughters of Missouri Rioter would have fallen 180,000 pounds behind their dams in the same time. Counting the milk at $1.50 per hundredweight, the income from thirty daughters of Lorne of Meridale would exceed that from the same number of daughters of Missouri Rioter $6467 in six years.

The next bull was Missouri Rioter 3d, a son of Missouri Rioter, and was the only valuable offspring the latter left in the herd. His mother was the best cow in the herd up to that time. Missouri Rioter has only three daughters with records of 15 lactation periods. Records are available for 14 lactation periods by the dams. The results are as follows:

	Dams	Daughters
Average yield of milk . .	4775	8005
Average per cent fat . . .	4.97	4.80
Average yield of fat . . .	238	384

The daughters produced on the average 3230 pounds of milk and 146 pounds of fat per year more than the dams. The number is of course rather small to admit general conclusions to be drawn with safety.

The least increase made by his daughters over the dams' records was an average of 1481 pounds of milk, and all three were of such outstanding quality that it is certain this bull was a remarkable breeder. Had the value of this bull been known, he could have made a fortune and a reputation for any breeder. He was raised on the college farm, and as his value was not recognized until too late, as has been the case with many of the greatest breeding animals, he was disposed of, and no record even kept as to what became of him.

Minette's Pedro was the next bull at the head of this herd. He was an animal of fine breeding, with many high-producing animals in his ancestry. This animal has had twenty daughters in the herd, with a total of 66 lactation periods.

	DAMS	DAUGHTERS
Average pounds of milk	5321	5376
Average per cent fat	5.04	5.04
Average yield fat	268	271

Ten daughters fall below their dams, and ten make some gain. On the whole the daughters are practically on a par with the dams, gaining only 55 pounds of milk and 4 pounds of fat per lactation period. It should be pointed out that the average of the dams is considerably higher than with the first bulls used, which sets a higher standard for comparison. On the whole, the herd was just holding its own with this bull at the head, and no general improvement was made. Some of the best as well as some of the poorest animals in the herd were daughters of this bull. Their dairy quality seemed to follow the dams in most cases, while with Missouri Rioter 3d and Lorne of Meridale the daughters were good regardless of the dams.

The last bull in the herd with daughters old enough to admit of a comparison of any value is Brown Bessie's Registrar. This bull has five daughters, with eight lactation periods. The comparison is made with the corresponding periods of the dams.

	DAMS	DAUGHTERS
Average pounds of milk	6029	4295
Average per cent fat	4.86	5.05
Average yield fat	293	217

The data are too limited to mean much as yet, but it is certain his daughters are decidedly inferior compared with the dams. The herd was deteriorating rapidly as long as he stood at the head.

These results show the immense difference in the way dairy qualities are transmitted, even where both sire and dam are pure-bred animals, and how serious a problem is the selection of a herd bull for the man who is trying to further improve a herd of pure-bred animals already well developed.

Cause of the Variation in Transmission of Dairy Qualities. — One of the chief difficulties in regard to selecting the bull is that practically nothing can be predicted from the looks of the animal as to how he will transmit dairy qualities. The man who will discover some means of so selecting the bull will confer a benefit on breeders that can scarcely be estimated.

There are two principles that are especially concerned with breeding, which should be kept in mind. The first is that "like produces like;" and the second is the law of "natural variations."

The cow, in the condition nature made her, undoubtedly produced only milk enough to feed the calf a few months until it could subsist on other foods. This milking characteristic was transmitted quite regularly. It was a case where like generally produced like; but some cows even then were undoubtedly better milkers than others, due to the law of natural variation. The principle of selection did not come in to retain these variations, and no improvement in this characteristic was made.

After cattle were domesticated, the same conditions existed, but finally man began taking advantage of the natural variations, and saving breeding stock from those having the characteristics, such as great milk production, which were found to be valuable. An animal which is different from others of its kind by natural variation will in some cases transmit the new characteristic to its descendants regularly. Such an individual is called a mutant, and is recognized as of the greatest importance in breeding operations. Other individuals may have desirable characters not common to their kind, but do not transmit these characteristics. In every breed animals have been found that transmitted characters in a remarkable manner. Some noted herds have been developed mostly from the progeny of a single animal. Among dairy cattle Stoke Pogis 3d and Hengerveld DeKol are good examples.

The breeder in selecting a bull for breeding purposes wants the animal that will transmit certain characters, and the question is how to select such an animal. What has been said explains the fact often learned by experience that in many cases the dairy qualities of a remarkable cow are not inherited by her offspring. The chances that the characters will be transmitted are greatly increased if found in the ancestry of both parents. This makes it important to select both parents carefully.

The rule of "like produces like" is only true to a limited extent; and the farther we get away from the original type, the smaller the proportion of cases where it holds good. This accounts for the fact often observed that the offspring of a phenomenal cow may be very ordinary. However, it will be found that on the average there will be more good animals among the offspring of such a cow than among those from a cow of moderate or low dairy capacity. We must always expect to find inferior animals appearing frequently in all herds. No breeder can prevent it; but no good breeder fails to reject the inferior ones promptly when discovered. The higher developed we get our cows, the more difficulty we must expect in keeping them all up to the standard.

Methods of selecting a Bull. — There are two courses open in selecting a herd bull. One is to buy a young bull on the strength of the records of his ancestors, and trust to luck to a certain extent that he will be one that will transmit the desirable characteristics of his ancestors to a high degree. As a rule, such a bull will do fairly well at least in transmitting these characteristics. For the owner of grade cattle or herds of low dairy capacity, this method of selection is reasonably satisfactory.

In selecting a young bull, the pedigree including the records of ancestors is of more importance than the individuality of the animal. The things to be looked for in the pedigree are first of all records of production by the female ancestors, especially the dam of the animal, and indications that the ancestors have transmitted the characteristics wanted.

There are some who refuse to have a bull from the phenomenal record-making cows for fear the vitality of the calf will be weakened. The majority of breeders, however, want the dam to have the highest record possible, other things being equal. We cannot expect more than a few of her close descendants will inherit this high quality but the chances are better for them to average up well than they would be from a cow of lower productive capacity.

There is a general belief among breeders that the characteristics of the dam of the sire are transmitted stronger to his daughters than are the characteristics of any other female ancestor. This view has not been demonstrated by conclusive evidence. Next in importance to the dam's record come the records of the sire's daughters. If this bull has sired many high-testing daughters, the chances are his sons will also transmit these characters.

Each animal inherits 50 per cent of its blood from its parents, 25 per cent from its grandparents, and $12\frac{1}{2}$ per cent from the third generation back. The relative value of the ancestors should be ranked according, and the common mistake avoided of attaching too much importance to an ancestor found in the third or fourth generation.

In studying records of production in the pedigree of a dairy animal, care should be taken to make certain what the records really mean. Official records mean more than private records, especially if the latter be churn records of butter produced. If butter records are given based upon a Babcock fat test, it should be noted what factor is used to make such estimate. The standard method of estimating butter from butter fat is by the addition of one sixth to the fat to represent the "overrun," that is the curd, salt, and water found with the fat in commercial butter. In some cases the butter production is expressed by adding one fourth to the fat, which gives a figure which is really too high. The most satisfactory method is to give pounds of milk and of fat. Records covering an entire year should receive much more attention than those covering only seven days. When the record covers only seven days and the fat test is abnormally high for the breed it is not safe to assume that this is a fair index of the richness of normal milk from the cow in question. The following is an example of the method of tabulating a pedigree. The records of production are usually written in red.

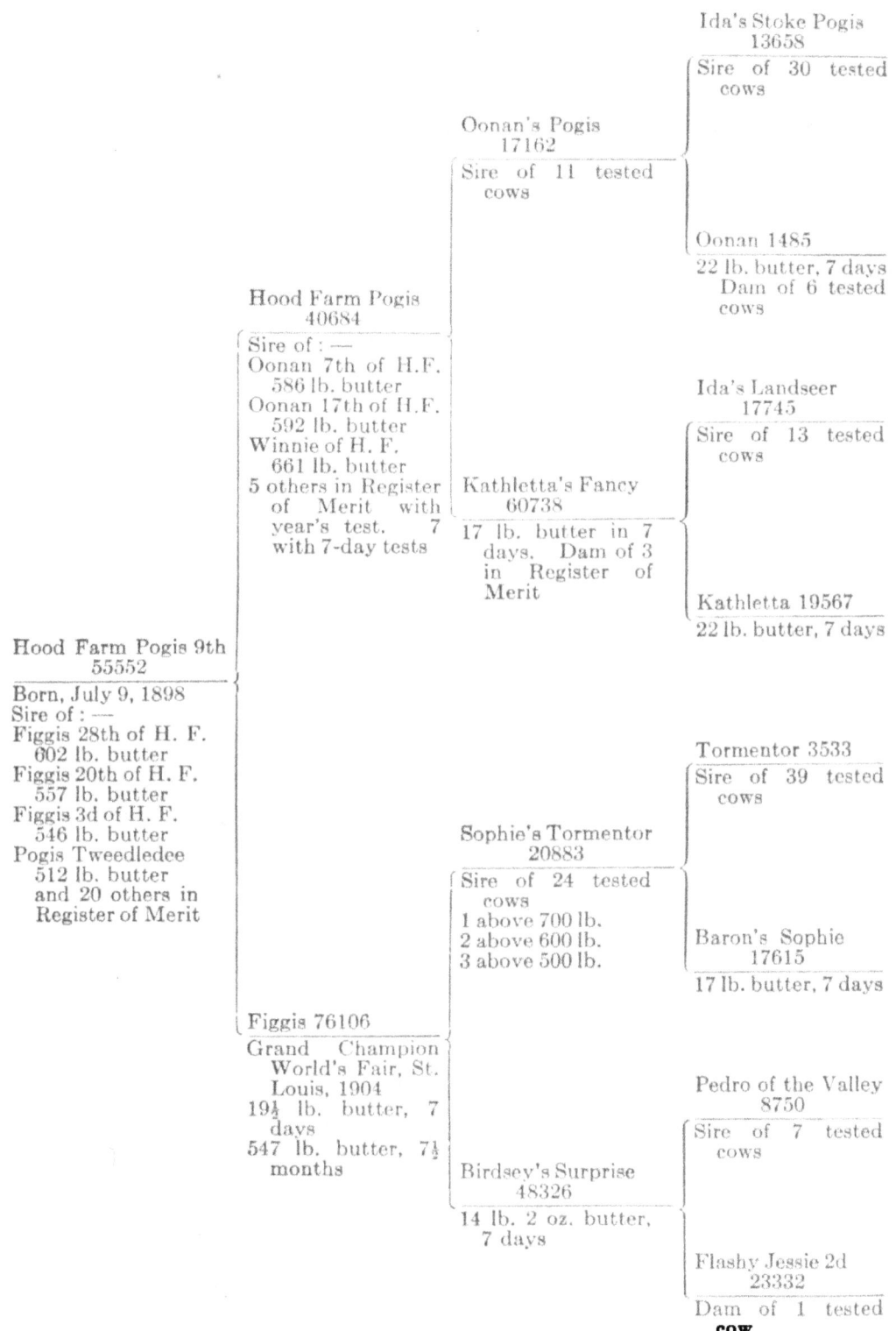

TABULATED PEDIGREE OF THE JERSEY BULL, HOOD FARM POGIS 9TH

The best means of becoming familiar with the leading lines of breeding in any breed is by reading the breed papers and studying sale catalogues and advertisements.

In buying a bull of any age, it should be required that he have a good conformation, strong vitality and constitution, and good breed characteristics. In buying a young bull the choice should fall upon one from a cow medium to large for the breed. She should be a regular breeder, and a cow of strong constitution and vitality. She should have a well-developed, symmetrical udder and teats, and a large year's milk and butter test, preferably official. While most dairymen favor the selection of a young bull as a herd bull, there always is the uncertainty about how he will transmit the dairy characteristics as pointed out.

A Tested Bull. — By all means the best plan of selecting a bull is to get one that has sired daughters of merit and showed himself to be the exceptional animal wanted by every breeder. The most skilled breeders are always on the outlook for such animals; but many are never discovered, and many others only after it is too late. Whenever possible, it is always advisable to retain an old bull until the results of his breeding can be ascertained. Then, if not satisfactory, the sooner he is gone the better; but there is always a chance of finding a bull like Missouri Rioter 3d, previously mentioned.

The wonderful prepotency of Stoke Pogis 3d was not recognized until after he had been sold for beef. Hengerveld De Kol, a well-known Holstein bull, on the other hand, was retained until it was discovered he was one of the great bulls of the breed, and continued in service until at present he stands first in the breed as the sire of tested daughters.

It is probable that in the future it will be more and more

common for the leading breeders to use a young bull in a limited way, then loan him or lease him to the owner of some grade herd until the results of his breeding can be ascertained. In this way the exceptional bulls that transmit the qualities desired can be found before it is too late. On the other hand, a valuable herd will not be damaged by the use of an animal that sires inferior animals. Essentially the above plan is now followed by the author with the University of Missouri Herd.

One of the unnecessary losses among dairymen is the constant sacrifice of bulls when mature and at their best. The average dairyman buys a young bull, uses him two years, and offers him for sale without waiting to learn of the quality of his daughters. His neighbor, instead of buying the old bull, buys a young one, and the older one that may be worth a fortune to the community is sold for beef while the neighbor is experimenting with the young one.

Age of Bull to Select. — It is an unsettled question whether an old bull is any more prepotent than a young. This claim is made by many, but with no real evidence upon which to base the claim. In Holland a breeding bull is seldom kept past three or four years of age. In this country a young bull is generally preferred by the average dairyman, mostly because a young one is easier handled and shipped than an aged one, and perhaps because on account of improper care the aged animal often is an uncertain breeder.

There is one danger connected with the aged bull that should be understood and guarded against. This is the possible introduction of contagious abortion into the herd. The greatest precautions should be taken on this point, and unless the buyer is satisfied the bull has not been contaminated with this disease, a young calf should be selected which is free from abortion, even if coming from a herd where the trouble exists.

Effect of Age of Dam upon Prepotency of the Bull. — The question of whether a bull which is the first calf of a heifer is as likely to transmit dairy qualities to the same degree as a son of the same cow after she is mature and has made large milk records, is one of importance that remains to be settled. It is believed by breeders of race horses that the racing qualities are transmitted better by mature and trained racing parents. At any rate, it is not wise to select the first calf of a heifer for a herd bull unless the calf is old enough so the heifer has had opportunity to demonstrate her dairy qualities.

CARE AND MANAGEMENT OF THE BULL

The bull calf designed for breeding purposes should be well fed from birth to maturity. The object in view in making such a recommendation is to allow the animal to develop to its limit. An underfed animal remains undersized, and while his progeny will not necessarily be smaller on account of a characteristic caused in this way, it is impossible to know whether his undersize has been caused by inheritance or underfeeding. For this reason a male breeding animal that is small and undeveloped for his age and breed is always looked upon unfavorably. Every male breeding animal should be handled in such a manner that it is certain he reaches full natural development.

A bull calf of a dairy breed is usually raised on skim milk, as the results are equally as good or better than those obtained when whole milk is used. A rather liberal grain ration should be fed. While there is no advantage in getting the bull fat while young, on the other hand, there is no special harm done if he appears a little smooth and beefy in form at this age, since this tendency will disappear later. The ration should be such that a good growth will be encouraged. As a winter ration, clover, alfalfa, or other leguminous hay is best adapted, while for grain a mixture of corn, with oats or bran or oil meal gives good results. Some breeders continue the feeding of skim milk until the animal is eight months or even a year in age, in order to insure rapid growth. The bull calf should be separated from the heifers at least by the time he is six months old, as he soon begins to annoy the heifers, and is better confined alone or with other bulls of his age.

Most bulls are sufficiently mature for light service at ten or eleven months of age. It is not advisable to breed one to over five or six cows before he is one year of age. From 12 to 15 months, not over one or two cows per week should be bred. As the animal becomes better developed, the amount of service may be increased. A well-fed, mature bull might serve 200 cows in a year, if distributed throughout the twelve months; but since the greater number usually are to be bred within a much shorter time, one bull is not usually expected to serve for more than a herd of fifty cows.

When the bull is about one year of age a ring should be placed in his nose as a convenience in handling. It also serves as something of a protection in case the animal becomes vicious and attacks the herdsman. In ringing a bull the animal should be tied securely. A bull punch sold by dealers in such supplies may be used for making the opening. The ring is slipped through as the punch is withdrawn. A trocar is equally satisfactory. In using it the instrument is forced through the cartilage division between the nostrils and withdrawn, leaving the cannula in the opening. One end of the opened ring is then passed through the opening, as the cannula is withdrawn. In case neither of the above instruments are on hand, an opening may be made with a sharp-pointed knife or with a 32-caliber wad cutter, an implement used in refilling gun cartridges. In using this a block of wood is placed on one side of the cartilage to be cut, and the hole cut by striking the cutter with a hammer. The ring should be filed or sandpapered smooth after it is closed, as rough edges often are formed by opening the ring or by tightening the screw.

The bull should not be allowed to run loose in the pasture with the herd. In the first place, it is a dangerous custom, as it gives abundant opportunity for the bull to attack a person unawares or where escape is difficult. Further, a record of the date of breeding of the cows cannot be kept, and no dairy herd can be handled to the best advantage without such records. Cows are often served younger or sooner after calving than the owner intends. The bull running with the herd exhausts himself until he becomes an uncertain breeder sooner than should be the case.

While nearly all successful breeders keep the bull confined, there is great need of an improvement in the way the bull is handled when confined. Too often he is confined in a dark, dirty stall without exercise from the time he is a calf. Such treatment is certain to result in weakening of the breeding powers, and often the animal becomes entirely impotent while scarcely more than mature. As a result of such treatment, by the time the sire is old enough to have daughters in milk so his value as a sire may be judged, he is too often valueless for breeding purposes. The main points to be observed in keeping a bull in good condition so as to retain his breeding powers are to avoid excessive use when young and to give plenty of exercise and a moderate ration when mature.

A box stall 10 by 12 feet with a ground floor is as suitable as any for such an animal, when he is to be kept in the barn. A strong paddock outside should be provided, with a door from the box stall. In all except severe climates the bull is best housed in a shed built in the paddock. All the protection necessary is a good roof and tight walls to serve as a windbreak. Exposure to any but severe weather is beneficial rather than injurious to the breeding bull. Such open-air treatment, with the precaution that the animal gets plenty of exercise, will keep the bull in the best breeding condition, and he will be as sure a breeder at eight or nine years as at two. This open-air treatment causes the animal to look somewhat rough, and for this reason does not find favor with those who make a business of selling breeding stock, since the appearance of the herd bull is of importance in making sales.

For those who prefer the herd bull to make the best appearance at all times, the best plan is to use a box stall in the barn, and see that the animal is outside only on fair days. Special attention should be given to exercising the bull under these conditions.

Where more than one bull is used, they may be kept together with advantage in a suitable paddock. There is no trouble or danger in keeping two or more bulls of any age together, if they be dehorned. One of the chief advantages is that they take more exercise than when confined alone.

One of the most common plans of exercising the bull when kept closely confined is by the use of a tread power. In some cases he operates the cream separator regularly, and occasionally other machinery. The power is adjusted to keep the bull walking at rather a slow pace as long as is wished. Another plan is to arrange a long sweep on a post and tie the bull at the end, and allow him to revolve the sweep by walking. Another plan followed is to arrange an overhead cable upon which a ring is placed from which a chain hangs. The bull is tied to this chain, and can walk the length of the cable.

There is some difference of opinion regarding the dehorning of the bull. Some breeders do not practice it, on account of a belief that the dehorned bull is less prepotent as a breeder. Another reason for not dehorning is that it injures the chances of the animal in the show ring. There seems to be no evidence, however, that dehorning does have the slightest effect on the prepotency of the animal. A bull is dangerous at all times, but more so with horns; and for this reason it is to be recommended that the horns be removed.

The bull of a dairy breed is more liable to be vicious than one of a beef, since the former are more active and have more nervous energy than the sluggish beef animals. In handling bulls it should always be taken for granted that they are dangerous. It is never safe to trust them, and it is usually one that has always been considered perfectly safe that injures some one.

A bull should be treated kindly, but never petted even when a calf, and anything even bordering on teasing must be prohibited. The bull is best let strictly alone except when it is necessary to handle him. He must always be handled in a firm manner and made to understand that a man has power to control him and must be respected. The man handling a bull must not show fear of the animal, and must take care that the bull does not gain the upper hand. The animal should be thoroughly trained to being tied and led when a calf. If this is done, he never forgets it, and may be tied or led at any time later, even if handled only at long intervals.

A plenty of exercise is one of the most important factors in preventing a bull from becoming vicious. It is also well to have his stall or paddock so located and constructed that he can see the other cattle and the attendants. Solitary confinement in an isolated box stall is not conducive toward the development of a quiet disposition in a bull.

Care should be taken that the bull never has a chance to learn to use his really enormous strength in breaking fences, gates, tie straps, etc. Keeping everything strong and in good repair will do much toward keeping the bull in subjection.

In leading a bull a staff should be used. In case he is regularly tied in the barn, a large, strong halter may be used, or a special bull stanchion, as preferred.

CHAPTER XIV

CALF RAISING

IMPORTANCE, RAISING ON SKIM MILK

Birth Weight of Calves. — The following table gives data compiled from the records of the University of Missouri dairy herd. The calves were weighed within 24 hours after birth, while the weights of the cows are the average of three days' weights, taken immediately following calving.

BIRTH WEIGHT OF CALVES

BREEDS	MALES		FEMALES		TOTAL FOR BOTH SEXES			
	No.	Av. Wt.	No.	Av. Wt.	No.	Av. Wt.	Av. Wt. Dam	Per Cent Wt. of Calf to Dam
Jerseys	35	56	29	46	62	53	864	6.24
Holstein	14	93	15	85	29	89	1127	7.87
Ayrshire					7	64	949	6.71
Dairy Shorthorn					8	76	1249	6.05

This table shows the males average heavier than the females at birth. Of the four breeds represented the Holstein calves are the largest at birth, and are also heavier in proportion to the weight of their dams.

More than three out of four calves come within 10 per

cent of the average weight. Breed is the largest factor influencing the size. The maturity of the cow also has some effect, as shown below. In this case cows are classed as immature with their first and second calves, and those that had given birth to three or more considered mature.

INFLUENCE OF MATURITY OF DAM UPON BIRTH WEIGHT OF CALVES

JERSEYS

	No. of Calves	Av. Wt. Dams	Av. Wt. Calves
Immature dams	23	791	50
Mature dams	39	906	55

HOLSTEINS

	No. of Calves	Av. Wt. Dams	Av. Wt. Calves
Immature dams	17	1004	82
Mature dams	12	1295	97

The size and vigor of a calf at birth is not influenced as much as might be expected by the previous feeding and condition of the dam. The reproductive function is so strong that in case of insufficient feed the fetus draws on the mother's body, and it is the mother that suffers mostly from the deficiency in nourishment. On the other hand, when a cow is excessively fat, the calf in many cases is somewhat undersized.

Importance of Raising Calves. — The success of the dairy farmer depends to no slight extent upon the careful rearing of the calves. The careful dairyman sees in every heifer calf the possibility of a cow that will not only replace a discarded member of his herd, but help to raise the average production. By proper care in the choice of the sire, and by careful attention to the rearing of the calves, the dairyman who is compelled to start with a herd of ordinary quality may, within a few years, raise the average of production of his herd to a marked extent. On the other hand, carelessness in breeding and in calf raising is bound to result disastrously to a herd, or at least to keep it at a standstill, as far as improvement is concerned.

One of the common mistakes made in the localities where whole milk is sold for market purposes or to condenseries or cheese factories is the failure to raise any calves. In this case the milk producer depends upon buying cows to replace those discarded from his herd. The excuse for this practice is that the cost of raising the cows is too great. Under such a system a dairyman will almost invariably produce milk year after year without improving the standard of his herd in the least. When more cows are required, they are purchased from a shipper or dealer, and without any information available regarding the merits of the animals beyond what can be determined from appearance. The dairy cows placed on the market through such channels are almost certain to be of very ordinary grade, since a cow whose value as a milk producer is known is not offered for sale at the market price.

In a few localities the practice is even worse, inasmuch as the cows are purchased when about to calve, and are milked but a single milking period, and are then fattened when the milk flow slackens, and sold for beef. In this way the general average of production is exceedingly low, and the occasional good cow is not even saved for future usefulness. The only means by which the average quality of dairy herds in the hands of practical dairymen can be materially improved is by the raising of their own cows by using a pure-bred sire and saving the heifers from the best cows.

Raising Calves by Hand. — The dairy calf is almost always reared by hand, although a small portion of the milk that goes to make the total production of the country is from cows whose calves are allowed to take a portion of the milk from their mothers. As a rule, the milk of the dairy cow is worth so much more than the calf that it receives the first consideration. A discussion of calf raising is naturally divided into two parts. One deals with calf raising where skim milk is available, as is generally the case when cream or butter is sold. The second part is concerned with methods of raising the calf where the whole milk is sold for market purposes or to a condensery or cheese factory. In this case no skim milk is usually on hand, and the calf must be fed whole milk, which makes an expensive ration, or be given some substitute for milk.

Raising Calves on Skim Milk. — It is a well-established fact that a calf can be raised on skim milk that is equally as good as one nursed by its mother. In localities familiar with dairying this is well understood, but in other places is virtually unknown. Some have seen unhealthy and undersized calves that have been fed skim milk, and have considered them as the necessary result of feeding skim milk. Such calves are the unfortunate victims of their owners' ignorance or carelessness. The skim milk calf raised according to modern methods differs little, if any, in size, quality, and thrift and value from the same animal when raised by the cow. The poor results which have so often followed the feeding of skim milk have been due to the faulty methods, and not to the fact that the cream which has been taken out is absolutely indispensable to the normal development of the calf.

The following table shows the average composition of whole milk and separator skim milk: —

	WHOLE MILK	SKIM MILK
Water	87.10	90.50
Fat	3.90	.10
Proteids	3.40	3.57
Sugar	4.75	4.95
Ash	.75	.78

It will be observed that the skim milk differs from the whole milk only in having most of the fat removed. The other constituents are slightly increased. The butter fat or cream is by no means the most valuable part of the milk for the calf. The fat does not go to form growth in a young animal, but to keep up the heat of the body and to supply fat for body tissue. The same material can be supplied much cheaper in the form of corn meal or other grain. The raising of the calf on skim milk is economical, because it is possible to make this substitution of a comparatively cheap grain for butter fat which has a commercial value for human food out of proportion to its food value for a calf. The parts of the milk which furnish the growth-making material are the casein and albumin, the former of which is seen as a white curd when milk is soured. From this material is made the muscles and bone, nerves, hair and hoofs, and this remains in the skim milk. The calf fed on skim milk is not generally so fat during the first six months of its life as the one nursed by the cow. It often has, however, rather a better development of bone and muscle, and the difference between the two cannot be seen two weeks after weaning time. The following suggestive figures from Storrs[1] Experiment Station show the results of feeding milk rich in fat as compared with milk having less fat.

SOLIDS FOR ONE POUND OF GAIN WITH CALVES

	MILK POOR IN FAT 3.27%		MILK RICH IN FAT 5.1%, 4.6%	
	Calf No. 1	Calf No. 2	Calf No. 3	Calf No. 4
Weight of calves at beginning of trial	66.5	45.0	67.0	58.5
Weight of calves at end of trial	133.5	79.0	126.0	94.0
Total gain of calves	67.0	34.0	59.0	35.5
Number of days fed	63.0	30.0	53.0	30.0
Gain of calf per day	1.06	1.13	1.11	1.18
Pounds of milk consumed	697.0	277.0	562.0	267.0
Pounds of milk solids consumed	777.9	30.9	78.4	35.6
Solids for one pound of gain	1.16	.91	1.33	1.03
Solids for 1 pound gain, average each pair, pounds	1.03		1.18	

The table shows that the milk containing the lower per cent of fat gave a larger gain per day and required a less amount of solids.

[1] Bulletin No. 31.

Feed required for a Skim Milk Calf the First 180 Days

	Storrs [1]		Missouri [2]		
No. animals ..	9	8	2 Spring Calves	2 Spring Calves	3 Fall Calves
Length of period	180 days	180 days	180 days	180 days	180 days
Breed	Dairy Breeds	Dairy Breeds	Jerseys	Holsteins	Jerseys
Weight at beginning ..	59	65	53	94	51
Pounds whole milk ...	90	220	334	400	367
Pounds skim milk ...	3001	2908	2422	3660	2331
Pounds hay .	337	618	46	115	275
Pounds grain .	127		111	69	159
Pasture ...			90 days	90 days	
Weight at end .	284	315	268	390	250
Average gain per day pounds .	1.25	1.31	1.19	1.64	1.10

The above table shows that the calf can get along with as little as 90 pounds of whole milk, although more is generally fed. The skim milk fed varies as a rule between 2300 and 3000 pounds, and, while calves can be reared without grain when skim milk is fed, it is customary to feed up to 150 pounds per animal during the first six months.

Fall calves need from 300 to 600 pounds of hay the first six months, depending upon how much other feed is given. Spring calves may be put on pasture and given no hay, or they grow equally as well if kept confined the first three or four months and fed hay.

[1] Annual Report Storrs Experiment Station, 1903.   [2] Unpublished data.

In the figures given the Jersey spring calves averaged 268 pounds at six months, while the fall calves averaged 250. The records for the same calves show that at one year of age the spring calves averaged 360 pounds, while the fall calves reached 440.

Comparison of Whole Milk and Skim Milk Calves. — The following table from the Kansas Experiment Station[1] gives the comparative gains made by three lots of ten calves each, fed on skim milk, on whole milk, and nursed by the dam.

Experiment	Number of Calves	Days Fed	Average Gain per Head	Daily Gain per Head	Cost per 100 Pounds Gain
Skim milk	10	154	233	1.51	$2.26
Whole milk	10	154	287	1.86	7.06
Running with dams	22	140	248	1.77	4.41

The calves nursed by the dams and those fed whole milk made a slightly better gain than those fed skim milk, but it was at a much greater expense, as shown in the table. The skim milk calves consumed 122 pounds of grain per 100 pounds gain, while the whole milk calves consumed 58 pounds grain and 31.8 pounds of fat in the milk. At this rate 100 pounds of grain was equivalent in feeding value to 48 pounds of fat. The economy of substituting the butter fat by grain is apparent, since the market value of butter fat is usually from twenty to thirty times that of corn meal fed to supplement skim milk. In the experiment quoted the calves which were steers were later put in the feed lot, and all fed for seven months. The best gains made were made by the skim milk

[1] Bulletin No. 126.

lot, followed by the whole milk lot, while the lot raised by the dams stood last.

Taking the Calf from its Mother. — There is some difference in practice regarding the time of beginning hand feeding. Some take the calf away from its mother at once without allowing it to nurse at all. Others prefer to let it nurse once; and some allow it to remain with the cow three or four days, or until the fever is out of the udder and the milk is fit for use in the dairy. It probably makes very little difference as to this point, but it is a fact easily established that the earlier the calf is taken from the cow, the easier it will be to teach it to drink. One that has never nursed is easily taught to take the milk from a pail, while one a week or a month old is often a difficult subject to teach. If the cow's udder is in good condition when the calf is dropped, it will generally be more satisfactory to take the calf away early. When the udder is caked, it is best to leave the calf with her until this condition is removed. One point that must be kept in mind is that at first the milk from the mother should always be given the calf, and not milk from some other cow. The first milk, or colostrum, given by a cow is especially suited to the requirements of a young calf, as it has the property of acting as a physic and stimulating the digestive organs.

Amount of Milk to Feed. — Under natural conditions the calf takes its milk frequently and in small quantities. The calf's stomach at this time is not suited for holding a large amount, and an excessive amount always results in indigestion and scours. For the first two weeks five or six quarts, or about ten or twelve pounds per day, is all the largest calf should be allowed to take. A small calf, as a Jersey, does not need over eight or ten pounds per day on the start. This may be fed in two feeds per day, or better, in three for two or three weeks. As the calf grows older, somewhat more milk can be used, but at no time does it need over sixteen or eighteen pounds or eight or nine quarts per day; but it is safe and economical to feed as high as twenty pounds to a large calf, if skim milk is plentiful. Overfeeding is undoubtedly one of the most common causes of inferior calves. It is a mistake to think that because the cream has been removed, the calf needs more of the skim milk, or that because the calf is not doing well it is not getting enough milk, and to allow it to gorge itself, which it will readily do, if given an opportunity. A good rule is to always keep the calf a little hungry. Some provision must be made for making certain that each animal gets its share and no more. Some drink twice as fast as others; and if fed together, one will be overfed and the other starved. The plan sometimes used of feeding a bunch of calves together in a long trough is very unsatisfactory for this reason, and should never be followed.

Temperature of the Milk. — Another precaution that must be taken is to have the milk warm and sweet when fed. Nature furnishes the milk to the calf in this condition, and we must carefully imitate her here. The digestion of a calf is quickly upset by feeding warm milk at one feed and cold milk at another. For the first few weeks the calf is especially sensitive to the temperature of its feed. After it is three months or more old, the milk may be fed somewhat cooler, if care be taken to have it at the same temperature all the time. Even then, however, the best results are secured when the milk is fed warm. The temperature of the milk should be that of the blood, or approximately 100° F. In this matter the feeder should exercise great care and not go by guess or by the feeling of the milk, but should actually use a thermometer often enough to know what blood heat feels like. If a hand separator is used, the milk may possibly be fed while still warm enough if used immediately after separation, but it will usually be necessary to heat it some artificially, if used for young calves during cold weather.

Changing to Skim Milk. — For the first two or three weeks the calf should be fed part of its mother's milk. However, in raising calves of those breeds producing a very rich milk, the calf will thrive better if the whole milk given during the first two or three weeks be diluted some with skim milk. Then the ration may be gradually changed to a skim milk ration by putting in a small amount of skim milk at first, and gradually increasing the amount day by day until at the end of a week all of the whole milk has been eliminated.

Supplements to Skim Milk. — The calf should be taught to eat grain as soon as it will take it. This it will generally do by the time it is three weeks, or, at most, a month old. The grain is best fed dry after feeding the milk. If the calf is with others, it will generally learn from them to eat when large enough. When the calf does not begin to grain as early as it should, it can be taught by putting a little grain in its mouth after the milk is drunk. In a few days it will begin to look for the meal, and will eat it, if offered in a box within reach. When once it begins eating grain, the calf is well started toward a good growth.

For the first few days grain should be kept before the calf. After that the ration given should be such that it will be eaten up clean each time. By the time the calf is six weeks old it usually will eat about one half pound of grain per day; at the end of two months one pound per day, and a month later two pounds per day. At no time up to six months is it necessary to feed more than this amount, although, if it is desired to push the calves along rapidly, they may be given more, up to the limit of their appetite.

There has been considerable investigation regarding the best supplement for skim milk. Since skim milk lacks in fat, it has generally been assumed that the supplement is especially to replace this fat.

The most extensive investigation along this line has been made by the Iowa Experiment Station. The following summary shows the results: —

GRAIN FOR CALVES WITH SKIM MILK

EXPERIMENT WITH SIXTEEN CALVES

	LOT I FOUR FED OIL MEAL	LOT II FOUR FED OATMEAL	LOT III FOUR FED CORN MEAL AND FLAXSEED	LOT IV FOUR FED CORN MEAL
	Lb.	Lb.	Lb.	Lb.
Milk	3760	3752	3760	3759
Hay	1478	1481	1478	1484
Oil meal	429			
Oatmeal		605		
Corn meal			538	601
Flaxseed			59	
Gain in 74 days	483	498	489	509
Average daily gain per head	1.63	1.68	1.65	1.72
Dry matter per pound of gain	4.13	4.31	4.32	4.16
Cost of feed per pound of gain	2.4¢	2.4¢	2¢	1.8¢

This experiment shows that satisfactory gains were made by using any of the rations fed, but that slightly the largest gains were made by using corn meal, while the dry matter per pound of gain was the lowest with oil meal. On account of the lower market value of corn, it made the cheapest gains. The conclusion from this experiment is that several of the common feeds may be used with success to supplement skim milk, and the one to be used will depend upon what is available, and the market value.

In the corn belt, where an abundance of corn is always available and usually cheaper than the other grains, it will be the most satisfactory to use, and it will be unnecessary to buy feeds not grown on the farm.

It was found by the Kansas Experiment Station that shelled corn gives equal, if not better, results than corn meal, after the calves are well started eating grain. As the calves approach weaning time, if corn is the grain ration fed, a change can be made with advantage to part oats, bran, or oil meal. Otherwise the ration may become too wide, and not contain sufficient amounts of growth-making nutrients.

Feeding Hay and Pasturing. — Calves will begin to eat hay, if it is put before them, about as soon as they will eat grain. For young calves timothy hay is often preferred to clover or alfalfa, as the young calf may eat more than it can properly digest of these palatable feeds. Further, they are rather too laxative, and help to produce scours, the most common difficulty in calf raising. When turned out to grass, the calves are as well supplied as can be with rough feed, but care should be taken to get them on grass gradually, so they will not get off their feed.

Importance of Sweet Milk. — In order to make a success of raising the calf on skim milk, the condition of the milk must be uniformly sweet. Probably nothing can be done that will produce indigestion and scours with more certainty than to feed sweet milk one day and sour the next. The younger the calf, the more sensitive it is on this point. After a calf is well started, it is possible to raise it on sour milk, provided the milk is fed in the same condition every day, but the results are not as satisfactory as with sweet milk. The Kansas Experiment Station [1] compared buttermilk with sweet milk by feeding ten calves on each. Those having skim milk gained an average of 2.02 pounds per day for 126 days, while the group fed buttermilk averaged 1.79 pounds per day. The latter group had less trouble from indigestion than those fed sweet milk.

The Creamery and the Skim Milk Calf. — The whole milk creamery has been the cause of much trouble in calf raising, on account of sour milk. Where the milk is hauled several miles in the hot sun, warmed to the proper temperature for separating, and then sent home at just the right temperature to sour most rapidly, it results in the milk being sour much of the time when received by the owners, especially during the hot weather. This has often been one common reason for poor success in raising calves, even where the creamery system is fairly well developed. Fortunately, a means has been devised to remedy this trouble to a great extent. Within recent years most of the creameries sterilize the skim milk — as the process is generally called, although it does not really sterilize the milk — before sending it back from the factory. This consists in heating it to at least 180° F., by using steam in most cases from the engine exhaust. The hot milk is put into cans, and taken home without cooling. This scalding checks the scouring, and such milk should remain

[1] Bulletin No. 126.

sweet until the following day, and if thoroughly cooled can be kept over Sunday. In this way the calf can be fed on sweet milk, and good results had with creamery skim milk. The Kansas Experiment Station raised calves on such milk that gained just as much and did as well as others fed milk separated with a hand separator and fed at once. When sterilized milk is first given to calves, they are apt to object somewhat; but in a few days they take it readily and do well. The general experience seems to show it is less apt to produce scours than is unheated milk. Every farmer who is patronizing a creamery and raising calves should insist that the skim milk returned to him be sterilized. The creameries find it of benefit to themselves and to their patrons to do this. Cans are found to be much easier kept clean when the skim milk is heated, and the condition of the milk when received at the factory is considerably improved on this account.

The Farm Separator and the Skim Milk Calf. — The rapid introduction of the hand separator is the feature of the times in the dairy industry. It has largely solved the question of getting good skim milk for calf raising, as well as having several other important advantages. Warm, sweet skim milk, separated within a few minutes after being drawn from the cow, is in the best possible condition for the calf, and by observing the points already mentioned and as practiced by the most successful dairymen, little trouble will be had in raising as good calves as are raised in any way. The majority of those producing cream or butter for sale insist on some means of raising the calf satisfactorily, and the hand separator seems to fill the want better than any other system.

Importance of keeping Pails Clean. — One common cause of sickness in hand-raised calves is feeding from dirty pails or cans. Every utensil which comes in contact with milk to be used for feeding should be kept clean and scalded as thoroughly as though the food were to be used for the owner's family. A good rule is to keep the calf pails as clean as the milk pails. In feeding grain, no more should be fed than will be eaten up clean. If grain is allowed to remain in a trough, it often becomes damp and partly decayed, and may cause sickness, just as dirty pails will often do.

Clean Pans and Barns a Necessity. — Another point to be kept in mind is that the young calf must be kept in a clean, well-bedded stall while in the barn. Experience has taught many men that a calf will not do well in a damp, dirty pen or stall. The calf needs all the sunlight it can get, and the well-lighted stall is always best. In arranging a barn the sunniest part should be reserved for the calf pens. In the summer the calf should have access to a small pasture with plenty of shade.

Plenty of Water Needed. — An abundance of clean water should be accessible at all times or at frequent intervals, as the calf is not satisfied with milk alone as a drink, and wants to drink a little water at a time, quite often during the day. This thirst for water is often overlooked when calves are raised by hand, and as a result the calf is thirsty as well as hungry, and gorges itself with milk when it has a chance. Salt should also be within reach when the calf is old enough to eat grain and hay.

Fall or Spring Calves. — There are a number of advantages in having calves to be raised by hand dropped in the fall. The calf can be kept growing nicely on skim milk until the grass comes, then weaned and turned out to pasture without checking its growth in the least. The disadvantages of winter feeding and cold weather are more than offset by the hot weather and annoyances from flies experienced by the spring calf. For the calf under six months it does not make much difference whether the roughness be grass or hay. Some prefer the latter, but for the second six months grass gives much better results. In the winter season the young calf is also more apt to get the careful attention it needs than it is during the busy summer season. As most heifers come into milk at about two years of age, a fall-dropped heifer is ready to begin milk at the season when the results are the most satisfactory.

CHAPTER XV

CALF RAISING

WITHOUT SKIM MILK, FEEDING WHEY, CALF MEALS, CALF FEEDERS, CALF SCOURS, VEAL PRODUCTION

Raising Calves without Skim Milk. — Where the whole milk is sold as market milk, or to a cheese factory or condensery, the problem of raising the calf is, how to do it without the feed costing more than the value of the animal raised. A calf needs about two gallons of milk per day in order to make good gains. This means 480 pounds per month, or 2400 pounds for five months. At the low price of $1 per hundredweight this would mean $4.80 per month, or $24 for five months, for milk alone to raise a calf under the best conditions. This excessive cost results in many cases in the dairyman adopting the plan of buying cows as needed and raising no calves. As already pointed out, this policy is certain to be disastrous.

Where calves are raised under these conditions, the plan most commonly adopted is to use the minimum amount of milk, using grain or some other substitute as far as possible. Some prefer to use cows that for some reason it is not desirable to milk, to raise the heifer calves. In this case it is usually possible for each cow to nurse at least two calves. Another plan is to feed the fresh milk rather freely for two or three months to start the calf growing, then remove the milk gradually and substitute something else, usually a grain mixture that supplies the nutrients needed in proportion somewhat similar to milk.

A variety of substances and mixtures have been tried for this purpose with more or less success. Grain mixtures in which oil meal and wheat middlings or flour constitute the major portion are most common. Calf meals prepared and sold under certain proprietary names are also used to some extent in this country, and quite generally in England.

The Pennsylvania Experiment Station [1] reports trials of two substitutes for milk prepared by them. The first consisted of: —

	Lb.
Wheat flour	30
Cocoanut meal	25
Nutrium	20
Oil meal	10
Dried blood	2

The second mixture consisted of: —

	Lb.
Corn	13
Nutrium	20
Flaxseed	1.5
Dried blood	2
Flour	30
Cocoanut meal	6
Oat chop	6

The best results were obtained with the first given.

The nutrium used in these mixtures was a soluble skim milk powder. In feeding the milk substitute it was mixed with warm water at the rate of one pound for six pounds of water, and fed from a bucket or calf feeder. The calves

[1] Bulletin No. 60.

were given their mothers' milk for from five to seven days, then the milk substitute gradually replaced the milk, until at the end of from ten days to two weeks no milk was given. For the first five or six weeks the calves were given about two pounds of the mixture per day; after that time two pounds and a half until weaning time. At the age of about 100 days the feeding of the substitutes was stopped, and these calves were put on a grain and hay ration. The thirteen calves raised on this ration consumed an average of 121 pounds of milk and 186 pounds of meal in 83 days, and the results, while not equal to those when milk was used, were satisfactory, and good dairy heifers were raised in this way at a low feed expense.

The ration used in this experiment was tried by the Massachusetts Experiment Station[1] in comparison with a commercial calf meal. Those fed the former gained 1.25 pounds per day up to six months of age, while those fed on the calf meal gained 1.15 pounds per day.

Whey as a Feed for Calves. — Where milk is used for cheese making, it is usual for the whey to belong to the milk producer, and it is used to a small extent as a calf feed. As a rule, the whey is fermented to such an extent before it reaches the farm that its use as a calf feed is out of the question. Where the cheese is made on the farm, or the whey is sterilized at the factory, it may be possible to secure it in a sweet, unfermented condition. Under the best conditions, it does not give results as a calf food that warrant it being recommended for this purpose. The average composition of whey, as compared with skim milk, is shown below: —

[1] Report Massachusetts Experiment Station, 1903.

	Skim Milk	Whey
Water	90.50	93.07
Fat	.10	.34
Proteids	3.57	.93
Sugar	4.95	5.00
Ash	.75	.60

It will be observed that the whey contains a little more fat, but only about one fourth as much proteids as skim milk. The sugar is a trifle higher. The removal of the greater part of the casein into the cheese takes out the most valuable portion of the milk from a food standpoint. If whey is used for calf raising, a grain ration should be selected that replaces as far as possible the constituents removed in the cheese. Corn is used with skim milk, since the latter is strong in proteids; but with whey conditions are different, and the proteids must be supplied. Oil meal is generally preferred for this purpose. About one half pound of oil meal is mixed thoroughly in a gallon of the sweet, warm whey, and fed as skim milk. If it is desirable to raise the calf on whey, it should be given its mother's milk for at least six weeks, and then can be changed to the whey. Some prefer to feed the grain dry. The other details of feeding and care are the same as given in regard to feeding skim milk.

Calf Meals. — A number of mixtures prepared and sold under proprietary names are on the market designed to take the place of milk in raising calves. These are usually a mixture of several ingredients, among which oil meal or ground flaxseed usually holds a prominent place.

According to the New Jersey Experiment Station [1] one of the most widely used calf meals is a complex mixture of oil-cake meal, bean meal, wheat middlings, cottonseed meal, carol beans, and fenugreek. The average composition of this meal, which seems to be quite uniform, as found from analyses made by nine experiment stations, is: protein, 25.2 per cent; fat, 4.90 per cent. Calf meals have been tested by a number of experiment stations. At the Ontario Agricultural College [2] it was found that calves fed calf meal as a substitute for milk in comparison with a ration of skim milk did better on the latter ration. Another calf meal was compared with ground oats and bran as a supplement to skim milk, but the best results were secured from the grain feeding.

Another milk substitute sold widely in Europe was tried in comparison with whole milk and skim milk by Moser and Kappele.[3] The mixture, which consisted mostly of oil-cake meal, bean meal, and corn, was found to cost one half as much as whole milk, and contained only one third the nutrients. Calves fed this ration gained .6 pounds per day, against 1.1 pound per day on whole milk. As a supplement for skim milk, the results were fairly satisfactory.

On the whole, the calf meals that are on the market give fair satisfaction as supplements to skim milk, but no better than grain mixtures that can be fed at much less expense. It is possible to use them as substitutes for milk in raising calves after they are two or three months old; but equally good results are found by feeding mixtures of grain pre-

[1] Bulletin No. 160.
[2] Ontario Agricultural College, Reports, 1900, 1905.
[3] *Landw. Jahrbuch d. Schweiz*, 1903.

pared by the feeder at less expense. The most satisfactory of the prepared calf meals seem to be those that contain a certain amount of dried skim milk.

Calf Feeders. — Several calf feeders have been devised and used in a small way. They are mostly made on the plan of having a rubber nipple for the calf to suck, and a tube of some kind to draw the milk from the pail. The claim is made that it is better for the calf to drink the milk slowly and mix the saliva with the milk which is done when the calf takes the milk from the rubber nipple. In practice it is found difficult to keep the tubes clean, and it is more work to feed the calf in this way than by teaching it to drink from the bucket. As to the injurious effects of rapid eating, there are no data at hand on the ground either way; but it is doubtful if it is of any special importance. Saliva probably has very little to do with the digestion of the milk, since the principal office of saliva in digestion is to moisten food and convert the starches into soluble sugars, and there is no starch in milk.

Ties for Calves. — Where valuable calves are raised, and it is desired to take every precaution to keep them in good condition, it is advisable to arrange a series of small pens so each animal may be kept by itself. This not only allows each animal to get the proper amount of feed, but enables the feeder to observe the individual more readily and detect any unusual conditions. A case of sickness may often be stopped by decreasing the feed of a certain calf after observing an abnormal condition of the manure so slight that it would not be possible to locate the affected animal in a group.

The next best arrangement, and the most commonly used, is stanchions. Where a group of calves run together, some means should be taken to tie them during eating. It is not only a great labor-saving device, but allows each calf to get its share of milk and grain. Calves should never be fed in a trough, as some will drink faster than others and be overfed, while others will be underfed. The same rule applies to the feeding of grain.

Stanchions for calves are made like ordinary rigid stanchions for cows, but smaller. A feed trough is put in front, with divisions to keep the feed of each calf separate. The pail of milk is set in the trough for the calf to drink. After drinking the milk, the proper amount of grain is put in the trough, and the calves left tied for some time to eat their portion. This usually prevents them from forming the habit of sucking each other, which is a point of some importance. If the calves are in the pasture, a convenient way is to fasten the stanchions on the fence.

Calf stanchions are usually made from thirty-six to forty-four inches high and twenty-eight inches from center to center, with a space of about four and one half inches for the neck. The feed trough should not be too wide; about fourteen inches is ample, with a depth of four inches where the stanchions are in the pasture and the calves are not to be fed hay. In the barn provision should be made for holding a sufficient supply of hay.

Scours in Calves. — The most common trouble in raising calves by hand is indigestion, or scours, as it is generally known. The chief causes of this common trouble, as already pointed out, are: overfeeding, sour or old milk, feeding cold milk, dirty pails, troughs, or stalls. The main thing to do is to prevent these troubles by keeping the conditions right all the time.

In considering scours, care should be taken first of all to distinguish between common scours, due to indigestion, and scours caused by navel infection.

White Scours, or Calf Cholera. — This trouble is quite common with calves. It usually appears within one or two days after birth. The calf is very sick from the beginning; the eyes become sunken, and a common symptom, although not always shown, is the passage of white, foul-smelling dung. The calf usually dies within a short time. This is a contagious germ disease, which gains access to the calf's body through the navel soon after birth. The freshly broken navel cord offers easy access to the system for the germs responsible for the disease. An animal once affected rarely recovers. If a calf has been affected with this disease in a herd, the chance of other cases developing is much greater. Often several cases occur in succession in a stall used for calving purposes. The trouble may be avoided by making sure that the calf is dropped in a clean stall, and that it is not allowed to come in contact with dirt or manure, until the cord is dry. If the herd is on pasture, it is best to allow the cow to remain there until the calf is born. Infection rarely occurs in the pasture. If the calf is born in the barn, the only safe plan is to tie up the cord at birth, and to apply a mild disinfectant, as a weak solution of creoline or zenoleum.

Scours from Indigestion. — This is often a serious matter in raising calves, as a bad case gives the calf a setback from which it recovers very slowly. Each animal should be watched closely for signs of indigestion. Often the first sign is foul-smelling dung. On the first indications of disorder the ration should be cut down to one half the usual amount. It is well to add one teaspoonful per pint of milk fed of a mixture of one half ounce of formaline in 15½ ounces of water. After two or three feeding periods, the milk given may again be increased to the usual quantity. The formaline should be given for two or three days at least.

When a severe case of scours appears, the feed should be at once reduced. A drench of three ounces of castor oil in a pint of milk may be then given with advantage. This may be followed by a teaspoonful of a mixture of one part salol and two parts subnitrate of bismuth three times daily two or three days until the condition of the animal improves. It is well to give the formaline mixture for several days while recovering from a severe attack.

Veal Production. — The production of veal is a question of minor importance at present, as far as American conditions are concerned. The veal supply of the American markets, as a rule, is irregular and inferior in quality. It is produced mostly as incidental to milk production, and not as a specialty, as is the case to a considerable extent in Europe. The chief supply for the city markets comes from the same herds that furnish the city with milk. When the whole milk is sold, most of the male calves, and too often the heifers, are sold for veal or disposed of in some way, and not raised.

The abundance and cheapness of other meats has resulted in a comparatively small demand for veal. As a result, the price has been low and the quality as a rule unsatisfactory. The dairyman selling market milk rightfully raises the question if he can afford to supply the milk necessary to feed the calf until it is of suitable age for veal. When cream or butter is sold, the margin is somewhat wider.

Feeding for Veal. — The calf designed for veal should be given all it will take of fresh whole milk if a first-class animal is to be produced. The calf is generally taught to drink from a pail, if it is to be kept very long. From eight to twelve pounds of milk at a feed twice a day is usually sufficient. To produce the finest quality of veal, the calf should not be allowed any feed other than whole milk. The veal is at its best at the age of two months. One put on the market under the legal limit is known as a "bob" veal, while one too old for good quality of veal is called a "heretic." The regulation of the United States government requires animals sold for veal to be not less than three weeks old. Many cities, and some states also, have regulations regarding the age or weight of calves that may be slaughtered. The age specified varies from three to six weeks. Where such regulations are not enforced, the tendency is to market the calves much younger than this, since where milk is high-priced, the younger the calf is sold the greater the profit.

Whether the calf can be raised for veal profitably or not depends largely upon the value of milk. It is often found that it costs more to feed calves, especially of the smaller breeds, than is received when the calf is sold, and for this reason many do not attempt to do so, but destroy all calves not needed for breeders at birth. Where a number of cows freshen at intervals, some of the calves are often raised that otherwise would not be by feeding the milk of the fresh cows up to the time it is fit for market.

The following data indicate what may be expected along this line: Hayward[1] fed six calves, weighing from 59 to 85 pounds at the beginning for 30 days. Average daily gain

[1] Annual Report, Pennsylvania Experiment Station, 1899.

was from 1.3 to 2.2 pounds. The calves consumed 393 pounds of milk on the average, containing 17.5 pounds of fat; and at prices then ranging were worth $4.73 each, an equivalent of $1.20 per hundred for the milk fed. A second test was made with eight calves, weighing from 58 to 80 pounds in the beginning. With a feeding period of a little over six weeks, these calves sold for a sum equivalent to 95.7 cents per hundred for milk; 9.8 pounds of milk testing 4.2 per cent fat were required per pound of gain.

The following data are presented from the Missouri Experiment Station: —

FEED CONSUMED AND GAIN FIRST 30 DAYS

	No. Calves	Average Birth Weight	Weight at End of 30 Days	Average Daily Gain	Lb. Milk per Lb. Gain
Jerseys	10	49	88.9	1.30	9.42
Holsteins . . .	8	83	127.0	1.46	10.02
Ayrshires . . .	2	70	107.0	1.23	9.31

These calves were fed whole milk, but were not fed to the limit, as might have been done had the production of veal been the object in view.

It will be observed from the data given that on the average one pound of gain on the young calf requires about 10 pounds of milk. When heavy feeding is practiced, this ratio seems to remain about the same.

If an abundance of skim milk is on hand, it may pay to feed calves for meat production to an older age before putting on the market and in this manner utilize the skim milk. Some dairymen using the dairy breeds find this fairly satisfactory, selling the calves at an age of from eight to ten months. However, the dairymen as a rule will find it more profitable to utilize the skim milk by feeding to growing pigs or laying hens than to raise this quality of meat.

Veal Production in Europe. — The farmers of Europe are in the lead in this line of meat production. There veal constitutes an important part of the meat supply, and the quality is excellent. The prices realized are sufficient to justify paying close attention to this line; and in some countries, especially Holland and Germany, it has become something of a specialty. In these countries where the finest quality of veal is produced, each calf is confined in a small pen, in a dark but comfortable stall, and is fed all the whole milk it will consume three times per day. They are put on the market between two and three months of age. The calves are said to consume about ten pounds of milk per pound of gain in live weight.

The following facts were given the writer by the manager of an estate in Denmark. The calves are fed whole milk for about two weeks, then gradually changed to a ration of 20 pounds daily of skim milk and one fourth pound sunflower seed cake. The last two weeks about one half whole milk is again fed, and the animals sold at an age of about twelve weeks at an average price of from $12 to $15 each.

CHAPTER XVI

THE DEVELOPMENT OF THE DAIRY HEIFER

FEED AND CARE AFTER WEANING. INFLUENCE OF FEED AND AGE OF BREEDING ON SIZE, CONFORMATION, AND DAIRY QUALITIES

Feed and Care of the Heifer after Weaning. — No difficulties are encountered in raising calves from the time of weaning until ready to come into milk. If the young animals are on pastures, no further attention is necessary, since grass furnishes the best and usually the cheapest growth.

The winter ration should consist of all the roughness the animals will consume, and a small amount of grain in addition. The object should be to keep the young animals in a thrifty growing condition without becoming unnecessarily fat. The liberal use of roughness is desirable, since it is usually the cheapest feed at hand, and further it is generally believed by experienced breeders that the consumption of large amounts of roughness when young helps to develop the organs of digestion to the maximum which is desirable when the cow comes into milk.

The roughness should by all means consist mostly of some legume, as clover, alfalfa, or cowpea hay, on account of the palatability and high protein and ash content of this class. Corn silage is also well adapted for part of the ration, but should always be combined with some leguminous hay or with a ration of grain that supplies ample material for growth, such as wheat, bran, or oats.

Age to Breed. — The age at which cows should come into milk depends somewhat on the breed and the maturity of the animal. The larger breeds, as the Holstein and Brown Swiss, as a rule should not calve much before thirty months. The more rapidly developing Jersey is, as a rule, sufficiently mature at two years. Other breeds rank in between these extremes. The proper age to breed depends somewhat on the size and development of the heifer. Heavy feeding of grain results in an animal large for its age and early sexual maturity.

Some breeders prefer to have the first calf dropped at rather a young age, claiming in this way to fix a habit of milk production in the young animal, and at the same time securing financial returns as early as possible. A common practice among these breeders is to allow about eighteen or twenty months between the first and second calves. In this way a long milking period is developed, and the cow has time to grow before the birth of the second calf.

It is a severe strain on the heifer to develop the fetus. Our investigations show that during the last three months of pregnancy a heifer, even when fed a liberal ration, does not add to her own body. The increase in weight shown is all found to be in the fetus. After the calf is born, the heifer weighs, as a rule, no more than she did three months previously.

Breeding too young undoubtedly results in small cows. It is impossible for a young cow to digest and assimilate a sufficient amount of feed to produce milk and growth at the same time. The production of milk, on account of its relation to reproduction, is a dominant function, and will not be materially checked to allow growth to continue. For this reason it can hardly be expected that a heifer calving young and immature in size will develop into a cow of normal size if she calves regularly each year afterwards.

The cow that has calved early as a rule shows a more pronounced feminine characteristic, and is finer in the bone, than the one that has calved at a later age.

The Development of the Dairy Heifer. — As already pointed out, one of the most important factors influencing economical milk production is the individuality of the cows used. It has been suggested that this probably is dependent mostly upon inheritance. Still the question arises, and is one of great importance as to the effect of the manner of raising the heifer upon the dairy characteristic when mature. In other words, is a superior or inferior cow born or made?

In 1906, with the view of getting data upon which to plan an investigation along this line, the Missouri Experiment Station sent a list of questions to the leading breeders of dairy cattle in the United States, to learn their opinion upon these factors that may influence the development of dairy heifers. Replies were received from 301 breeders, representing an experience of breeding 150,000 animals. The widest possible divergence of opinion was found respecting some of the points covered. A series of investigations was then undertaken to gather more accurate data under experimental conditions. According to the plan, one group, numbering 20 animals, was to be fed on whole milk, and to receive all the grain they would consume from birth until they came in milk. Another 20 were to be raised on skim milk and alfalfa or clover hay without receiving any grain until they came into milk. It was planned to have one half of each of these groups calve at 20 to 24 months of age, and the other half at 32 to 36 months. The conditions are extreme in both directions, in order to give some definite results in regard to the influence of the ration fed and of the age of calving. While this investigation is not yet completed, sufficient data are at hand to enable some conclusions to be drawn.

In discussing this subject, certain of the questions sent to the breeders will be given, followed by a summary of their opinions and a discussion based upon our experimental results.

Influence of Overfeeding when Young. — *Question 2.* Do you believe from your experience that a dairy cow may be injured by being allowed to become overfat when young?

Of 281 breeders replying to this question 76 per cent reply in the affirmative and 24 per cent in the negative. It has been believed for a long time by many breeders that it is injurious to the milking qualities of a dairy cow to become fat when young. This has been thought to develop a tendency toward using feed for body fat that will persist when the animal is mature and in milk. This supposition is based upon observations which may easily be erroneous. When a cow of a dairy breed lacks dairy qualities and shows a beef tendency in conformation, it is easy to attribute it to improper feeding when young. In many cases the same animal did show a beef tendency when young, but not from overfeeding, but as an inherited characteristic. So far in our investigations no injurious effect on the milk-producing function has been found from heavy feeding when young. Some of our best cows have been in one group, and some in the other. Those heifers which have been kept fat from birth until coming in milk lose the surplus body fat within a short time after calving, and show no more tendency to fatten later while in milk than do those raised on the light ration. The most marked effect of the heavy grain ration, as found so far, is a much more rapid growth and a quicker maturity as compared with light-fed animals. However, the results show a heavy grain ration for dairy heifers is entirely too expensive as compared with a ration made up mostly of good roughness. From an economical standpoint the feeding of a heavy grain ration cannot be practiced.

Corn in the Heifer's Ration.—*Question 3.* Are there any feeds that should be especially avoided in raising dairy heifers?

Of 301 breeders answering this question 10 per cent answer in the negative; 48 per cent mention corn as a feed to be avoided; 10 per cent more avoid "fatty foods," most of these probably having feeds rich in carbohydrates, such as corn, in mind; 10 per cent also avoid cottonseed meal. Our investigations do not give any data on this question, except in so far that we have had splendid dairy cows that were fed excessively on a grain ration consisting of two thirds corn continuously up to the time of calving. It seems, however, that there is a widespread belief that corn is detrimental to the development of a dairy heifer.

Forbes[1] found experimentally that an exclusive corn ration for young swine caused a retarded development of proteid and bony tissues and overdevelopment of fat tissue. It resulted in fine-boned, poorly muscled, undersized, overfat animals, with impaired powers for breeding. It is reasonable to expect somewhat similar results would follow excessive feeding of corn to a growing heifer. However, this must

[1] Bulletin 213, Ohio Experiment Station, p. 302.

not be interpreted to mean corn is not a safe food. There is no time in the life of a dairy cow where it cannot be fed, if used with moderation and in combination with other suitable foods.

Factors Influencing the Size. — *Question 5.* How would you proceed if you wished to develop especially large animals?

Under this head we found 18 per cent of the breeders mentioned the selection of large parents, 82 per cent liberal feeding, and 17 per cent late calving as the factors to be taken into account. Our investigation so far bears out the breeders' opinions. We find liberal feeding of grain when young not only causes a much more rapid growth and development, but makes a somewhat larger animal in the end. The growing period of the light-fed heifer is longer, and if they do not come into milk until three years of age, they are not undersized. However, if they calve at 20 to 24 months, and thereafter at intervals of twelve months, they remain somewhat undersized. We find the character of the ration fed and the age of calving both to be factors influencing the size of the animal when mature. Continued selection of large parents would unquestionably tend to increase the size. However, according to the principles of heredity as now understood, parents made larger or smaller than normal by the food received will not transmit this characteristic to their young. If a cow or bull is undersized by inheritance, however, it is probable the offspring will inherit the same tendency.

Bulky Rations as Influencing Digestive Capacity. — *Question 7.* Does your observation indicate that the liberal feeding of roughness while young helps to develop a strong digestion?

Of 287 breeders replying to this question, 75 per cent answered in the affirmative. This indicates a widespread belief in the favorable influence of a bulky ration while young upon the digestion of the mature animal. Our investigations give no positive data along this line. It has been observed, however, that a heifer that has been receiving a heavy grain ration up to calving time will consume, when first put upon a typical dairy ration, only about one half as much roughness as a heifer that has been raised upon roughness. However, this marked difference soon disappears. After about two months on the same ration, we have been unable to observe any difference in the power of digestion. However, more accurate means must be employed to determine this point definitely. The digestive apparatus has great powers of adapting itself to the character of the food consumed, and for this reason it is probable that the effects of the ration fed during the period of growth are not permanent.

Age at first Bulling. — *Question 12.* At what age, on the average, do the heifers first come in heat?

The influence of breed counts for considerable in this connection. The average age given for Jerseys was 8 months, Guernseys 11 months, Holsteins 11.4 months, and Ayrshires 12.8 months. Our investigation has shown a wide variation in this respect due to the ration fed. As might reasonably be expected, the animals receiving a liberal grain ration reach sexual maturity sooner than those on a ration with no grain. Our data show an average difference for the two groups of 65 days for the Jerseys, 126 days for the Holsteins, 100 days for the Ayrshires; an average of 92 days for the 32 animals included.

CHAPTER XVII

MANAGEMENT OF DAIRY CATTLE

Dehorning. — For the ordinary business herd there is every reason why the animals should be dehorned. Before domestication horns were useful as a protection against other wild animals, but horns on a dairy cow of to-day serve no useful purpose, and are the source of much annoyance. They are responsible for frequent injuries, often serious, and especially to the udder. Dehorned cattle may be housed in a much smaller space; and when they are fed and watered together in the ordinary manner, there is a material saving in labor. Horns on a bull are extremely dangerous. The only case where it is advisable not to remove the horns is with high-class animals that are likely to be used for show purposes. While the scale of points for the various breeds allow only one or two points for horns, a dehorned animal loses much more than that score card indicates when in the show ring. Dehorned animals are occasionally found in the show ring, but the leading show animals are practically all horned.

Dehorning may be done on the grown animal by the use of the saw or clipper, or on the calf with caustic potash. When using the saw or clippers, the animal should be at least one year of age, and preferably two, or there is danger that scurs will later develop. This will always happen when a young animal is dehorned unless the horns are properly removed; that is, cut sufficiently close to the head. The horns should be cut from a quarter to one half inch below where the skin joins the horns, leaving a rim of the skin on the horn removed.

The best method of dehorning is to use caustic potash on the young calf. To use this successfully it must be done before the calf is more than three days old. The hair is clipped away from the small buttons which may be felt, and which are the future horns. A stick of caustic potash is then moistened a trifle and rubbed on the spot until the skin bleeds slightly. Care must be taken to avoid getting too much water on the animal, or it may run down the head, taking off the hair, and even getting into the eyes, with serious results. If sufficient caustic potash has been applied, a dent will be felt in the skull after a few days, and no horns will ever develop.

Dehorning of grown animals is best done in cool weather of spring or autumn. If done in hot weather, means must be taken to keep flies out of the wound. In raising pure-bred animals for sale it is advisable to leave the horns on, as they can be removed, if the buyer wishes.

Directions regarding methods of tying animals for dehorning with the saw or clippers and details about the work may be found in Farmers' Bulletin, No. 350, and in the 24th Annual Report of the Bureau of Animal Industry, U. S. Dept. of Agriculture.

Marking Cattle. — In breeding pure-bred cattle it is very essential that some practical system of marking the individuals be adopted. Even if a breeder feels that he can depend upon memory alone, it is not advisable to do so. The records of some valuable herds have been lost through the death of the owner, who was the only one who knew the individual animals.

The common practice of taking dairy-bred calves away from the dams soon after birth to be raised by hand makes it much more uncertain to depend upon memory to identify these animals, than is the case with beef-bred cows where the calf is raised by the mother in the natural way and the owner has several months' time to learn the individuals.

When a buyer visits a herd, a much better impression is made if every animal bears a tag. Every calf should be tagged in some manner within a few days after birth, and a record made in the herd book to identify the animal. It is important, even in grade herds, especially when large, to have a permanent mark of some kind that may be used as a means of identification if question of ownership should arise.

There are several methods in more or less general use for this purpose. A common plan is the use of ear tags of various forms, bearing numbers, and, if desired, the name and address of the owner. Some of the forms in use are shown in Fig. 42. The chief objection to these is that they are frequently torn from the ear by being caught in a fence or branch of a tree, and in this way not only is the mark lost, but the animal disfigured as well. An advantage of this type is that it may be removed and another substituted if desired. In inserting these tags care should be taken not to close them too tightly on the ear, as a soreness may result which leads the animal to rub the ear and tear out the tag.

Various systems of notching the ear to indicate numbers are also in use. These are satisfactory, except that the animal is disfigured, and the system must be known to read the number. Another method in use is to burn the number on the horn or hoof. A mark on the hoof must be renewed at intervals, and one on the horn is occasionally lost by the horn being broken off, as well as being impractical in many herds on account of the practice of dehorning. The young calf cannot be marked in this manner. Some breeders mark the animals by placing a brass tag bearing a number on a strap about the neck. This is especially well adapted for calves.

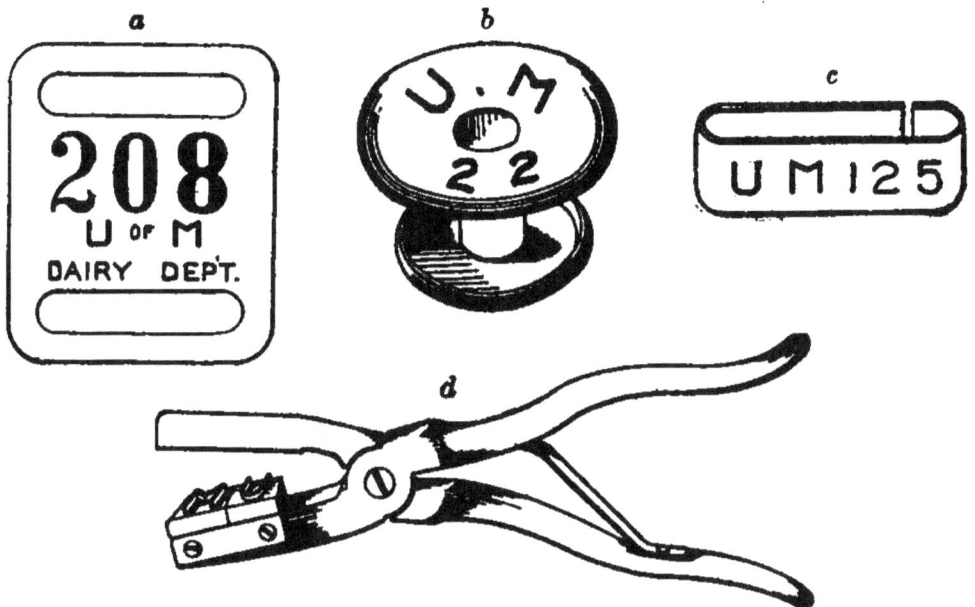

Fig. 42. — Devices for marking cattle. *a*, brass tag to go on strap; *b, c*, ear tags; *d*, tattoo marker.

For use with all breeds having light-colored skin in the ears the tattoo mark is the most satisfactory of all. This consists in printing the numbers or letters as desired in the skin of the ear with india ink. The instrument for making the punctures is arranged so any combination of letters or figures may be used. In using this the skin of the ear should be thoroughly cleaned of grease by washing with soap or gasoline. The puncture is then made, and the tattoo oil thoroughly rubbed in. The figures show clearly after a few days, when the wound is healed, and remain permanently.

The plan adopted by the author is to mark every calf before being taken from its mother. Each animal is given a permanent herd number, which is recorded, and a strap placed around its neck bearing this number on a brass tag. This tag is worn until the animal is nearly mature and familiar to the attendants. The tattoo mark is put in the ear at the age of about one year, and while not suitable for identification at a distance, makes it possible at any time to positively identify the animal.

PROTECTION FROM FLIES IN SUMMER

Decline in Milk Production in Midsummer. — In the latter part of the summer the production of milk by the average dairy herd falls off rapidly. At the same time, the animals are annoyed greatly by the flies, and in the popular mind they are looked upon as the chief cause of this decline. The rapid falling off in milk production is illustrated by a compilation made by the author.

Sixty farmers supplying milk to the college creamery at Ames, Iowa, sold only 46 pounds on August 1 for each 100 pounds delivered by the same parties on June 1, a decline of 54 per cent. At the same time, for every 100 pounds produced by the college herd on June 1, 68 pounds were produced on August 1. The actual difference was really greater than these figures indicate, since some fresh cows were added during the period to the farmers' herds, while the figures from the college herd represent only those animals that remained in milk during this time.

The large decline in production with the farmers' cows

is typical for the summer season under conditions where the dairy cow is made secondary to other lines of farming. Where dairy farming as a specialty is practiced, such great declines in production do not ordinarily occur, on account of better management. The much smaller decline with the college herd was the result of supplementing the pasture with green feed.

Flies that trouble Cows. — Cattle in this country are troubled mostly by two varieties of flies, known as the stable fly, *Stomoxys calcitrans*, and the horn fly, *Hæmatobia serrata*. The stable fly resembles the ordinary house fly in appearance; but while the house fly does not bite, the stable fly has mouth parts that enable it to pierce the skin and suck the blood of animals. The eggs are laid mostly in manure, horse manure preferably, but also in cow manure. The period for development is about fifteen days from egg to adult fly.

The horn fly was introduced into America about 1886. It is considerably smaller than the house fly, and gets its name from the characteristic of gathering about the base of the horns. It is also recognized by the habit of feeding with the wings spread, and usually goes in swarms. Its bite is very irritating, and causes a congestion resembling the bite of a mosquito. The eggs are laid in fresh manure, and require about ten days to develop adult flies. These flies remain with the cattle constantly, roosting largely on the horns.

Since the flies that annoy cattle are hatched in manure, the first precaution to be observed in reducing the numbers to the minimum is to avoid an accumulation of manure where it will remain moist, and especially near the barn. Since horse manure seems to be preferred by these pests, special care should be taken not to allow it to remain in heaps near the barn. Where a small amount of manure accumulates, it is sometimes kept in a screened inclosure. When the weather is damp so the manure dropped in the field remains moist a sufficient length of time, the flies hatch freely wherever droppings are found. In dry seasons this opportunity does not occur, and the number of flies is usually much smaller.

The popular idea of the great injury done by flies has resulted in many proprietary mixtures being put on the market designed for the purpose of keeping off the flies. The compounds are usually composed chiefly of some coal tar product with the addition of fish oil, resin, or pine tar. These are applied to the animal with a hand spray pump. Great claims are made by the manufacturers regarding the injury done by flies and the profit resulting from using these repellents. Investigations made by at least three experiment stations have failed to show any advantages from their use.

Lindsay[1] reports trials with ten brands of fly removers. Four of these he found were efficient in keeping off the flies, the others were not. He did not determine the effect on the milk and butter production of the cows.

Beach and Clark[2] report an extensive trial to determine the effect of applying one of these preparations on the milk and butter fat yield of cows. The trial covered two summers. They concluded that while the preparation they used protected the cows fairly well against the flies, the milk and fat production was not increased when the cows were sprayed. Their investigation indicates that the annoyance of cows by flies is overestimated.

The author[3] carried on a similar trial during the summers of

[1] Fifteenth Annual Report, Mass. Experiment Station.
[2] Bulletin 32, Storrs Experiment Station, Storrs, Conn.
[3] Bulletin 68, Missouri Experiment Station.

1903 and 1904. The first year sixteen and the second year twenty-one cows were used. The fly season was divided into periods of two weeks, and the entire herd sprayed each morning on alternate periods. When the fly repellent was applied each morning, the cows were fairly well protected during the day. The only advantage found was that the cows were less restless during milking. No effect could be detected upon the yield either of the milk or of the fat.

The plan has been recommended by some of housing the cows during the day in a darkened barn to avoid annoyance from flies, letting them out at night to graze. The objection to this is the difficulty of keeping the stable dark and at the same time cool and comfortable. Again, the extra expense of labor in cleaning the stable and supplying bedding makes it impracticable in most cases.

The main cause for the marked falling off in milk during the summer is to be attributed to the failure on the part of the cows to eat a sufficient amount of food. The excessive heat in a warm climate has more to do with this than the flies, although the latter may contribute somewhat to the general effect.

The pastures are often short also at this time, making it more difficult for the animals to gather sufficient feed for the heaviest production. It will be observed that during hot weather the cows will graze but little during the day, and come to the barn at night evidently hungry. While the influence of these conditions cannot be entirely removed, they may be improved. The first thing is to make certain the cows do not lack for food. They should be in the pasture during the night if possible, and at least during the coolest part of the day, during early morning and late evening. The feeding of silage at this season is also to be recommended to supplement the pastures.

Sheltering. — The housing of the dairy cow naturally depends upon the climatic conditions. She should not be exposed to severe weather, while cold rains and snowstorms are especially to be avoided. The most favorable temperature has not as yet been determined experimentally, but observation indicates that a barn temperature of between 40° and 50° F. is as favorable as any. During the winter season the cow should remain in the barn, except for a few hours in the middle of the day when the weather is mild. On stormy days or during periods of excessively cold weather she will do better if kept inside constantly. It is well to provide an arrangement to supply water to the animals in the barn in cold climates.

An abundance of fresh air is as necessary for the health of the cow as of any other animal, and should be provided without fail. However, it should be supplied by proper ventilation, and not through the walls of poorly constructed barns. Excessively warm weather is far more injurious to the dairy cow than cold, and there is no practical means of making the animal comfortable under such conditions. For this reason hot weather and warm climates are not conducive to a high production of milk. This is especially the case where high temperature is combined with a high humidity of the atmosphere.

MILKING

Milking the Heifer. — If the heifer is properly handled before she has her first calf, there is little difficulty in teaching her to be milked. If the heifer has been tied while being fed as a calf, there will be no further trouble about tying her at any time. The heifer should be accustomed to the stable where she is to be tied for some time before she calves. It is best to tie her for a month or more before she freshens in the stall she is to occupy when in milk, and to make a point of handling her daily. A careful man should have the milking of her at first, and must go about it carefully and without exciting her. Very little trouble will ever be experienced under such conditions. The men who care for cows should always move about among them gently and not startle them by sudden movements or loud talking.

Methods of Milking. — Milking is generally considered such a simple operation that any common laborer is supposed to be able to milk. However, there is an immense difference in milkers, and one of the most difficult parts of carrying on dairy farming is securing competent men to do this work. One milker may be able to get 20 per cent more milk than another, one may dry the cow within a few months, while another may keep her in milk the entire year. The milker should not be allowed to excite or worry the cows by loud talking or cruelty or abuse of any kind.

The secretion of milk is involuntary, but may be affected indirectly by excitement of any kind. Even the presence of a stranger or a dog at milking time is sufficient to affect the milk yield of many cows. The changing of milkers results in some loss for a few milkings, unless the new milker is better than the former.

A cow should be milked quietly and quickly. A cow is largely a creature of habit. If usually fed at the time of milking, she cannot be milked satisfactorily until she has her feed. Special care should be taken to secure all of the strippings. The first milk drawn may contain as little as 1 per cent

220   *DAIRY CATTLE AND MILK PRODUCTION*

of fat, while the last drawn runs from 6 to 10 per cent. In milking the whole hand should be used, closing first that part next to the udder; then the milk is forced past the sphincter muscle by closing the remainder of the hand. The milking should not be done by using the thumb and forefinger alone,

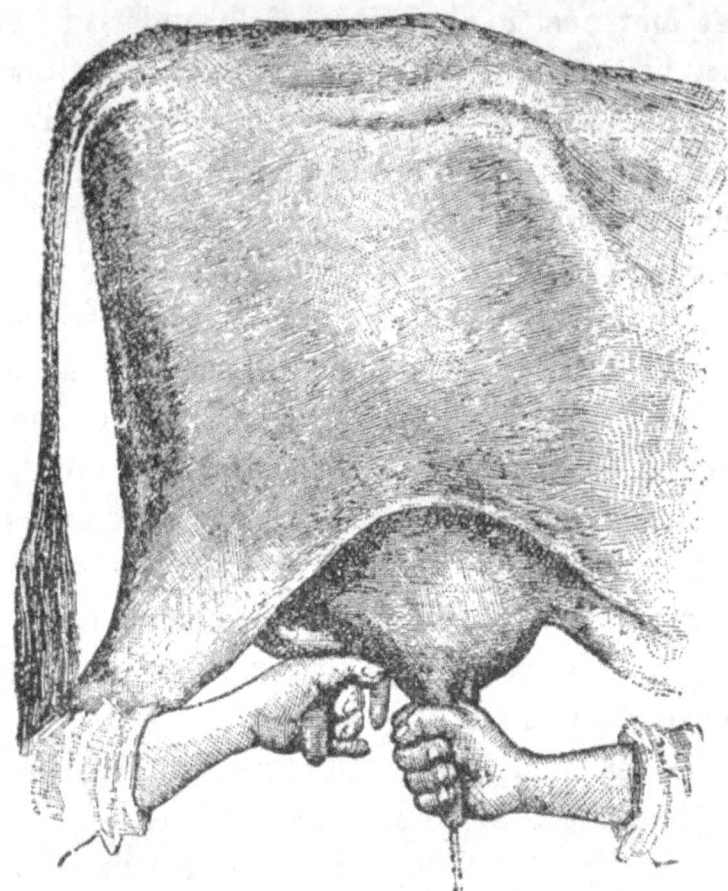

Fig. 43. — Correct position of the hands when milking (Grotenfelt).

neither should the thumb be inclosed within the palm, as is sometimes done. The cow's teats should always be dry when milking. Wetting the teats not only is a dirty, filthy practice, but it also allows the teats to chap and become sore in cold weather. If there is difficulty in milking dry, a small amount of vaseline may be rubbed on the hands. This serves the same purpose as wetting the teats, and is beneficial rather than harmful, both in a sanitary way and as it affects the cow's teats.

Effect of Period between Milkings. — If a cow is milked twice a day at twelve-hour intervals, there is as a rule little difference between night's and morning's yield or richness. If the period is unequal, the larger amount of milk and the lowest quality usually follows the longer period.

Milking three times per day is practiced with heavy-producing cows, and with all cows that are being crowded for the largest records, especially if it be for short periods. Few cows can produce over 60 pounds of milk per day with two milkings, and when 75 to 80 pounds are reached, the production will seldom go higher unless the cow is milked four times each 24 hours. When the udder becomes congested to a certain point, no further secretion takes place until this congestion is removed by milking. With heavy-producing cows it will pay in a practical way to milk three times daily. With cows of anything like ordinary productive capacity, the increase is not sufficient to pay for the extra labor involved. The richness of the milk is also somewhat increased with heavy milkers by milking more than twice per day.

Stoppage of the Teats. — It occasionally happens that the opening in the teats is apparently closed up when the cow begins to produce milk. Something must be done to open it, or that portion of the udder will be spoiled. This may often be done with a common silver milk tube or with a teat expander. Small, hard lumps form occasionally at the base of the teat, and should be treated with the bistoury.

The Milking Machine. — A satisfactory milking machine has long been the greatest need of the dairy farmer. While still in the experimental stage, the milking machine is at present sufficiently well developed to be a commercial success in the hands of some operators. It is thoroughly demonstrated that it will milk cows, and that by its use a skillful operator can do as good work as the average milker. It seems, however, that it is not yet equal to a good milker in the amount of milk that will be secured. The cow is not injured in the least, and in fact prefers machine to hand milking. Most users have found it more difficult to hold up the yield of milk with the machine through the entire lactation period than when hand milking is practiced. If the machine is properly cleaned and used, the sanitary condition of the milk is much improved over any ordinary conditions; but with careless handling the quality of the milk may be even worse than hand milking. So far it is only practical for herds of 30 cows or more.

The Hegelund Manipulation. — This system of manipulation was originated by Dr. J. Hegelund, a teacher in a Danish agricultural school in 1900. It was brought to the attention of dairymen in this country by Professor Woll of Wisconsin. It is really a manipulation of the udder made by the milker as soon as the regular flow of milk has ceased, in order to secure all the milk secreted by the cow. This manipulation serves the same purpose as ordinary stripping, but is claimed to be more efficient. The system has been tested by Woll[1] and by Wing and Foord.[2] The manipulations, three in number, are described as follows by Woll.

Description of the Hegelund Manipulation. — *First Manipulation.* — The right quarters of the udder are pressed against each other (if the udder is very large, only one quarter at a

---

[1] Bulletin 96, Wisconsin Experiment Station.
[2] Bulletin 213, Cornell Experiment Station.

time is taken), with the left hand on the hind quarter and the right hand in front on the fore quarter, the thumbs being placed on the outside of the udder and the four fingers pressed toward each other and at the same time lifted toward the body of the cow. This pressing and lifting is repeated three times, the milk collected in the milk cistern is then milked out, and the manipulation repeated until no more milk is obtained in this way, when the left quarters are treated in the same manner.

Second Manipulation. — The glands are pressed together from the side. The fore quarters are milked each by itself by placing one hand, with fingers spread, on the outside of the quarter and the other hand in the division between the right and left fore quarters; the hands are pressed against each other and the teat then milked. When no more milk is obtained by this manipulation, the hind quarters are milked by placing a hand on the outside of each quarter, likewise with fingers spread and turned upward, but with the thumb just in front of the hind quarter. The hands are lifted, and grasp into the gland from behind and from the side, after which they are lowered to draw the milk. The manipulation is repeated until no more milk is obtained.

Third Manipulation. — The fore teats are grasped with partly closed hands and lifted with a push toward the body of the cow, both at the same time, by which method the glands are pressed between the hands and the body; the milk is drawn after each three pushes. When the fore teats are emptied, the hind teats are milked in the same manner.

Wing and Foord found that from 3 to 13 pounds, and an average of 8.75 pounds, of milk containing .63 pound of fat were secured per week from each cow by following the Hegelund Manipulation. It also appeared that this milk was in addition to the amount secured when the manipulation was not practiced. This would indicate the amount of this residual milk is all gain.

They further compared the Hegelund Manipulation with ordinary stripping. They conclude that, while the Hegelund Method is effective as a means of securing all the milk possible, practically as good results are obtained by ordinary stripping when properly carried out. These results are in accord with those found by Woll.

In one of their tests two milkers were leaving 20 cents worth of fat each every milking, that could be removed by proper methods. In another herd 8 per cent of the fat and 16.9 per cent of the fat was lost by failure to milk out all secreted. These investigations showed clearly the necessity of careful work in milking; otherwise heavy financial loss will occur. All the milk secreted by the cow should be removed, as that left in the udder by a careless milker is lost.

Hard-Milking Cows. — Some cows are considerable of an annoyance to the milker on account of milking unusually hard. This is generally caused by a strong sphincter muscle, which closes the teat opening tighter than it should. This condition may be remedied by proper treatment. For most cases the use of teat plugs is sufficient. These plugs, which are made of lead or hard rubber, are placed in the teat duct, and a tape tied around the teat through the eyehole in the plug. The animal wears the plug from one milking to another. This is continued until the muscles are somewhat relaxed and the opening remains larger. In some cases this treatment is not sufficient to cause a permanent enlarging of the opening. In other cases the difficulty in milking is caused by a hard lump in the upper part of the teat, which seems to drop down into the opening when drawing the milk. In this latter case, or when the use of an ordinary dilating plug is not sufficient, the bistoury should be employed. This is an instrument that is passed into the teat opening, and then by means of a screw a small knife edge is projected on one side at the upper end, and the instrument is then withdrawn. The knife blade cuts the side of the teat duct and the surrounding muscles. An or-

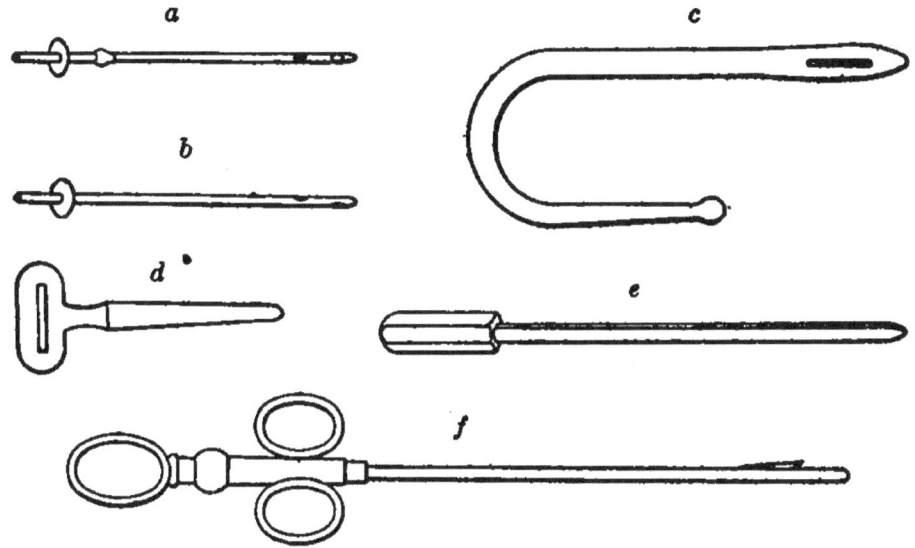

FIG. 44. — Instruments for treatment of udder troubles. *a, b*, milk tubes; *c*, lead teat plug; *d*, teat expander; *e*, teat opener; *f*, teat slitter or bistoury.

dinary teat plug is then kept in the opening, except at time of milking, until the cut heals. This leaves the opening enlarged permanently.

Sterilizing Instruments. — Before a milk tube, a teat plug, the bistoury, or any instrument is inserted in the opening of a cow's teat, it must be thoroughly disinfected. If this is not done, germs will gain access to the udder, which will cause inflammation that may ruin the animal. Instruments should first be thoroughly cleaned in warm water, and preferably boiled. Before being inserted into the teat, they should be placed in a 5 per cent solution of carbolic acid, or in a rather strong solution of creolin, and should be inserted while wet with this solution and without being touched with the hand on the portion that enters the duct.

Cows with Leaking Teats. — Some cows lose a portion of their milk by leaking from the udder between milkings, on account of the sphincter muscle having a weak contraction. No practical remedy has been devised for this trouble. Under conditions that warrant the small amount of trouble involved, the teat opening may be closed after each milking with collodion.

Bloody Milk. — Blood in milk is more common than is generally understood. It may be often noticed in separator slime when its presence was not suspected in the milk. It is not an indication of disease or any unhealthy condition in the cow. It is caused by the rupture of a small blood vessel that allows blood to escape into the milk cistern or the milk ducts. In some instances certain cows have it at intervals for a number of months, but more often it appears but once or twice. It cannot be prevented or stopped, and the only thing to do is to reject the milk affected.

Chapped Teats. — Sore teats may be caused by cold weather, milking with wet hands, or other causes of local irritation. When so affected the cow does not stand quietly for milking on account of the pain. The trouble may be easily remedied. The application of vaseline for a few times on the first appearance of the trouble will usually check and cure it. If severe the teats should be thoroughly washed and softened with warm water, after which glycerite of tannin may be applied. An application of equal parts of spermaceti and oil of sweet almonds is also recommended.

Warts on the Teats. — These are often troublesome. They often disappear or are greatly benefited by applying vaseline or olive oil. If large, they may be cut off with a sharp pair of scissors and the spot touched with a stick of caustic potash.

Bitter Milk. — It is not uncommon for the milk of certain cows to have an abnormal taste and smell when far along in the lactation period. This trouble is usually experienced, or at least most often attracts attention, where one or two cows are kept as a family milk supply. As far as the author has observed, this abnormal condition of the milk occurs only after the cow has been in milk seven months or more, and usually when she is far advanced in pregnancy. It only rarely occurs when the cow is receiving green food. The milk has a peculiar taste, described by some as salty, but more often as bitter. The author has observed that the abnormal taste is present in the fresh milk, but rapidly grows worse for 24 hours, regardless of the temperature at which it is kept, indicating that the growth of bacteria cannot be the cause. In most cases the animal was fat and receiving more feed than necessary when the trouble occurred. Reducing the grain ration to the amount actually needed by the cow, and giving two or three doses of Epsom salts, 1 to 1½ pounds at a dose, at intervals of three days removed the abnormal condition in some cases. Cream from milk in this condition churns with great difficulty, and sometimes cannot be churned by any method that can be devised.

Kicking Cows. — Cows are given to few vices, and those are mainly due to faulty management. The most common is kicking when milked. The cows always kick at first, either from pain or fear. If not handled properly, it may grow into a habit. Under no circumstances should a man strike a cow that kicks. It does no good, and always makes them worse. Gentle measures, however, will not work with all cows, and some old cows that have been taught to kick by mismanagement cannot be cured by the best of care. Such animals should always be tied during the milking. This is best done by taking a rather heavy strap with a strong loop. The strap is put around one leg above the hock, and the end drawn through the loop. The strap is then put around the other leg and buckled, so the two legs are held close together. The cow may struggle a little at first, but soon learns to stand quietly as long as the strap is in place.

Self-Sucking Cows. — This vice is not very common, but annoying when begun, and difficult to stop. If an ordinary cow contracts the habit, the best advice is to sell her at once. The most effective method of treatment seems to be to put a bull ring in the cow's nose and hang a second ring from the first. This method was suggested by the Wisconsin Experiment Station, and has been tried by the author with good success.

CHAPTER XVIII

MANAGEMENT OF DAIRY CATTLE (*Continued*)

Shall the Cow be given a Rest? — In practically every large dairy herd the common practice is to have all cows dry for a short period before calving. Others, more especially less experienced dairymen, make a practice of milking their cows continuously. The objection to this practice is that the cow needs a period in which to recuperate. The production of a liberal amount of milk is as much of a tax on the cow as heavy work is on a horse. The cow will produce more milk if dry six weeks than she will if milked continuously. It is often expected that milking up to the birth of the calf will result in the calf being weak and small from lack of proper nourishment. The author's observations do not bear out this statement. The mother and not the fetus seems to suffer most in case of insufficient feed.

A cow that is not given a rest before calving will begin at a much lower level of production than will be the case when she has had opportunity to recuperate, and this means a lower level throughout the lactation period. Under ordinary conditions the cow should be dry six weeks; and if she is in a thin condition, it is better to make it two months.

Drying up the Cow. — Where certain cows are milked continuously, it is sometimes claimed they cannot be dried up. There is little difficulty about this if properly handled. The common method of drying a cow is to lengthen the interval between milking by omitting one milking each day. After a few days the milk is drawn only once in two days, until secretion is completely stopped. This may require two weeks or more. There is far less danger of injuring the cow's udder in drying her up than is generally believed. If a cow is producing as little as 10 pounds per day milking can be stopped at any time, and no harm will result. The udder should not be milked out at all. It will fill up for a few days, and then the milk contained is gradually reabsorbed, and no harm will result in any case. If a cow is producing more than this amount of milk, it is advisable to first cut down her feed. The grain ration should be all removed, and if the cow is producing as much as 14 or 15 pounds per day, feed her only on timothy hay for a few days until the production of milk begins to decline, then stop the milking. The author has practiced this method for several years with high-producing cows, with no injurious results.

An example will show how it works in practice. In 1907 the Jersey cow, Bessie Bates, under the charge of the author, was producing 19 pounds of milk per day, testing 5.5 per cent of fat, and it was desired to dry her on account of approaching calving. Her grain ration was first of all entirely taken away, and she was given timothy hay only. After six days her production had declined to 14.1 pounds of milk per day, and at this point milking ceased entirely. Her udder filled rather full; but after three or four days it began to soften, and within a week was perfectly normal. In her next milking period of 365 days she produced 13,895 pounds of milk and 680 pounds of butter fat.

Management of the Cow when Dry. — The cow should be in good flesh at the time of calving, as discussed in other places.

This is important in order to start the cow at a high level of milk production. Further, far less trouble is experienced in parturition when the animal is in good order, and there is less trouble from retention of the afterbirth.

The feeding of the cow while dry will depend upon her condition of flesh when milking ceases. If she is in good order, that is, somewhat more than moderate in flesh, a little more than a good maintenance ration will be needed while dry. There is no more suitable ration for such a cow than good pasture or clover hay and corn silage. If she is thin, the ration should be sufficient to get her into proper condition at calving time.

The amount of nutrients needed for developing the fetus has not been determined, but must be taken into account. A calf usually weighs from 50 to 90 pounds at birth, depending upon the breed. No analysis of newborn calves is available, but it is probable they contain at least 60 per cent of water. On this basis a calf weighing 60 pounds would contain 36 pounds of dry matter. It would require 300 pounds of ordinary milk containing 12 per cent of solids to equal the dry matter in a calf of this size. On this basis it would require 12 days for a cow producing 25 pounds of milk per day to produce as much dry matter in the milk as is contained in a calf at birth. We cannot assume, however, that the tax on the animal would be the same in both cases.

At this time the cow should have exercise, and nothing is better in this respect than freedom in a smooth pasture. She should not be chased by dogs or driven through narrow gates. As parturition approaches, she should be put on a laxative ration if during the period of winter feeding; if on pasture no special attention need be given to the feed. She should not be tied in a barn where the platform inclines toward the rear. The average length of pregnancy is 285 days. The males are carried a little longer on the average than the females.

Should Cow be milked before Calving. — It is the practice of some to milk out heavy milkers several times before the birth of the calf, for fear the udder may be injured. It is also claimed by some that this helps to prevent milk fever. Others never milk a cow until the calf is dropped. If milking is begun it must be continued regularly. Milking before calving is advisable only with the heaviest milkers, when they are evidently suffering greatly from the distention of the udder.

Developing Long Milking Period. — While persistency in milking is mostly a breed and individual character, it is generally believed it is possible to influence the length of time a cow will give milk each period during her life by the length of the first milking period. For this reason it is best to milk a heifer in her first lactation period as long as she should be milked when mature, even if the milk produced does not justify the time, in order to establish the habit.

Care of the Cow at Calving Time. — If the cow has been dry for six weeks and received sufficient feed so that she is in good condition at calving time, there seldom will any complications arise. If the cow is on pasture, she should be allowed to remain there, but looked after at least twice per day when about to calve. If not on pasture, the cow should be turned loose several days before she is expected to calve in a box stall of sufficient size. Special attention must be given to avoid infecting the navel of the calf and bringing on contagious or white scours (p. 198). As the time of parturition approaches, the udder becomes distended and hard, and filled with the colostrum milk. When the tendons and muscles relax on either side of the rump, leaving a hollow appearance on either side of the tail head, parturition may be expected within 24 hours, or three or four days at the longest.

The cow should be left strictly alone at time of calving, unless some assistance is evidently necessary. As a rule the calf will be born within half an hour. If the calf is not expelled after an hour or two, an examination should be made. The normal position of the calf at the time of delivery is forefeet first with the front of the hoofs and knees upward while the nose lies between the knees. If the condition of the calf is normal, the cow may be assisted by pulling on a rope attached to the forefeet of the calf. This must be done carefully, and only when the cow strains. If the position of the calf is abnormal, the services of a qualified veterinarian should be secured if possible, unless the person in charge has had considerable experience. No attempt will be made here to describe abnormal presentations of the calf and how they are to be handled. The reader is referred to veterinary textbooks, and to the book entitled " Diseases of Cattle," issued by the U. S. Department of Agriculture, for full details on this subject.

The cow is especially subject to retention of the afterbirth, and special attention must always be given that it comes away. When the cow is in good condition, the afterbirth is usually expelled within a few hours after the calf, often almost immediately. Cows far along in years or in low condition of health are especially subject to this trouble. The giving of cold water soon after calving may cause it to be retained. All water given within the first 24 hours should be warmed, and cold feed should also be avoided. The afterbirth when expelled should be removed, to prevent the cow from following her instinct and eating it, which may result in disorders in the alimentary canal. If the afterbirth is not expelled, a serious condition of the cow is brought about by the decomposition of the tissues within the body and the absorption of the poisons. A cow in such condition becomes emaciated and produces but little milk, and that is not in fit condition for food. This condition is easily known by the fetid products that escape, and the offensive odors that may penetrate the entire barn.

The cow should be so handled that retention of the afterbirth will be prevented as far as possible. However, it will occur frequently in all herds. If it does not come away within 24 hours, it should be removed by the hand. There is no drug that can be used for the purpose. If taken in time, a weight of one or two pounds tied to the protruding membrane may by its dragging effect pull the membranes and stimulate the uterus to contraction. The only treatment that can be relied upon is to remove it by the hand. For the inexperienced the service of the veterinarian should be secured. Every man having the responsibility of caring for many cows should acquire the experience necessary to do this successfully himself. The following description of the method is given by Dr. Law.[1]

" The operation should be undertaken within twenty-four hours after calving, since later the mouth of the womb may be so closed that it becomes difficult to introduce the hand. The operator should smear his arms with carbolized lard or vaseline to protect them against infection, and particularly in delayed cases with putrid membranes. An assistant holds the tail to one side, while the operator seizes the hanging afterbirth with the left hand, while he introduces the right

[1] *Diseases of Cattle*, p. 218. Published by U. S. Dept. Agric.

along the right side of the vagina and womb, letting the membranes slide through his palm until he reaches the first cotyledon to which they remain adherent. In case no such connection is within reach, gentle traction is made on the membranes with the left hand until the deeper parts of the womb are brought within reach and the attachments to the cotyledons can be reached. Then the soft projection of the membrane, which is attached to the firm, fungus-shaped cotyledon on the inner surface of the womb, is seized by the little finger, and the other fingers and thumb are closed on it so as to tear it out from its connections. To explain this it is only necessary to say that the projection from the membrane is covered by soft conical processes, which are received into cavities of a corresponding size on the summit of the firm mushroom-shaped cotyledon growing from the inner surface of the womb. To draw upon the former, therefore, is to extract its soft villous processes from within the follicles or cavities of the other. If it is at times difficult to start this extraction, it may be necessary to get the fingernail inserted between the two, and once started the finger may be pushed on, lifting all the villi in turn out of their cavities. This process of separating the cotyledons must be carefully conducted, one after another, until the last has been detached and the afterbirth comes freely out of the passages."

Care of Cow after Calving. — It should be borne in mind that the vitality of the cow is low following parturition, and she should be treated accordingly. She should be protected from cold drafts, and in case of severe cold weather it is well to cover her with a light blanket for a day or two. The water given should be warmed for two or three days. The ration for the first few days should be light in character and not very abundant. A bran mash made by moistening bran with warm water is well adapted for the grain portion. With this can be given such amount of hay as will be readily eaten. If the udder is swollen and hard, the grain ration should be increased very slowly until this condition disappears, when more feed can be added, using as a rule two weeks at least, and with a heavy milker still longer, to get her on to full feed. No alarm need be felt if the udder remains inflamed and somewhat hard for a number of days, provided milk can be drawn from each quarter. The calf may be left with the cow for two or three days, or removed earlier, as desired. In either, the cow should be milked at least three times daily, or oftener, until the inflammation leaves the udder.

Milk Fever. — Until recent years the owner of high-producing cows always had to face the danger of losing the most valuable cows by this common and usually fatal disease. One of the most important discoveries of recent years for the dairy cow owner is the discovery of the air treatment, which is so simple any one of intelligence can apply it, and with almost certain success.

Milk fever occurs only with high-producing dairy cows. It never occurs with the first calf of a heifer, and seldom with the second. It affects mature cows, and especially the heaviest milkers. The well-nourished cow is more subject to it than the underfed, which was the reason for the practice, common before the present treatment was discovered, of withholding feed for several days before parturition. The great advance made in recent years in the records of milk production is to be attributed to a considerable extent to the air treatment, making it possible to have cows in a high state of flesh at calving without loss from milk fever.

The disease is so typical it is easily recognized. It occurs in nearly every case within 48 hours after calving, and usually only after normal parturition. Every cow liable to be affected should be watched carefully for symptoms until the danger is past. The first indications are restlessness and excitement on the part of the cow. Within a short time paralysis of the hind legs begins, resulting in a staggering gait. The animal soon falls and is unable to rise. From this time on the cow becomes unconscious, and remains so until death occurs in from 18 to 48 hours unless treated. The cow assumes a characteristic position, which is of great value in diagnosing the case. The head is turned to one side, and rests on the chest with the muzzle pointing toward the flank. The entire body is paralyzed, making it impossible to give medicine; but fortunately none is required.

The first effective treatment was discovered by Schmidt of Denmark, who injected a solution of iodide of potash into the udder. Later Anderson, also of Denmark, found that the injection of ordinary air is far superior to the first treatment. This is the method now universally used.

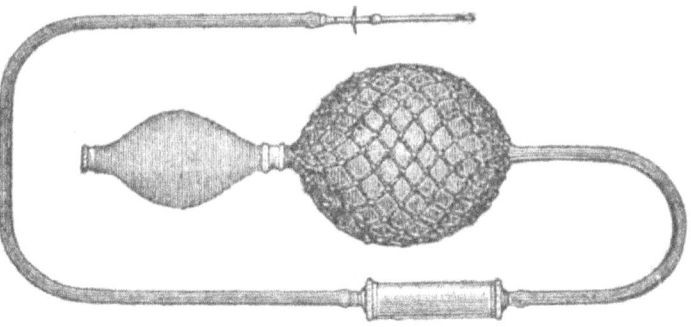

FIG. 45. — Apparatus for treating milk fever.

Every dairyman should be provided with a suitable milk fever outfit to treat cases promptly as they appear. The apparatus used may be of various forms.

238   DAIRY CATTLE AND MILK PRODUCTION

The most approved is shown in Fig. 45. The essential parts are a milk tube, to which is attached a rubber tube, a receptacle of some kind in which clean cotton is placed to catch the dust in the air as it is pumped through it, and a

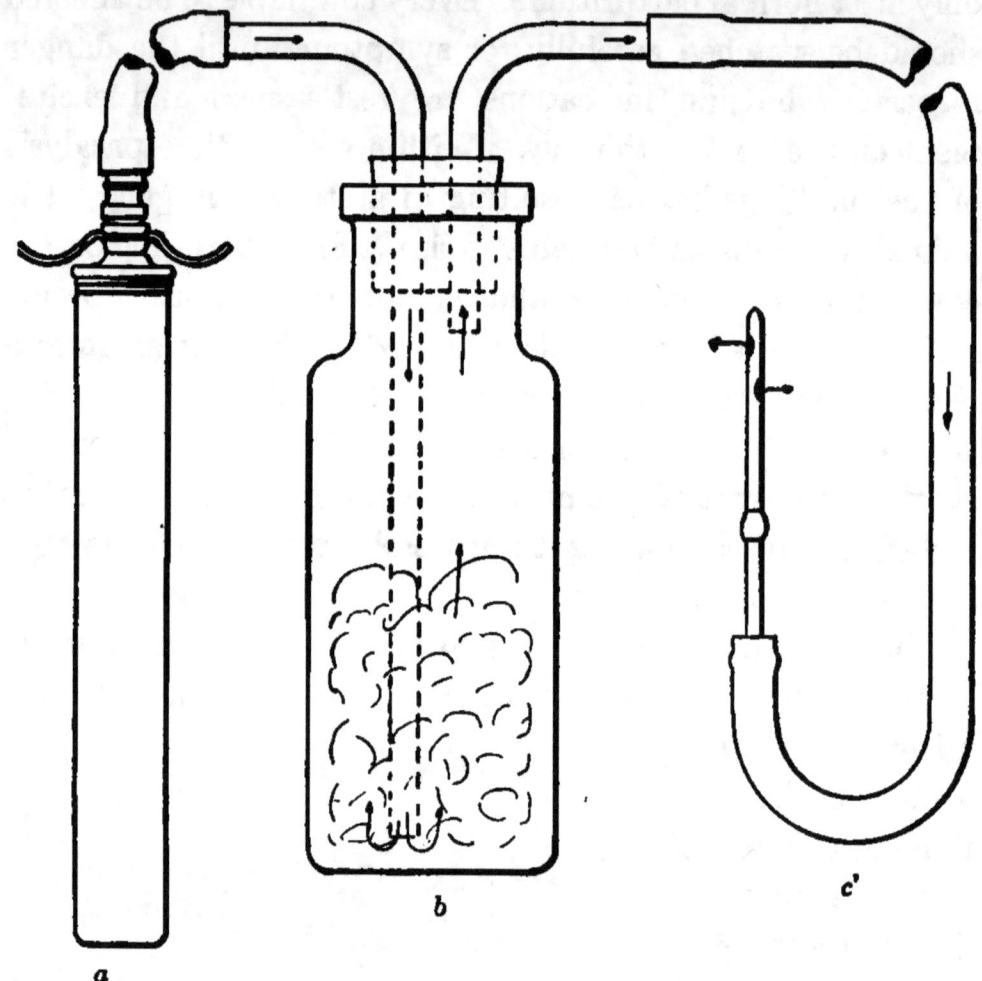

Fig. 46. — Improvised apparatus for treating milk fever. *a*, bicycle pump; *b*, bottle containing cotton; *c*, rubber tubing with milk tube at end.

rubber bulb or a pump of some kind. In case an approved form of apparatus cannot be secured, an apparatus can be improvised that will serve the purpose. Fig. 46 shows an apparatus used for several years by the author, which can be put together in almost any drug store. The author has known cases where a common bicycle pump with a quill for a milk tube was used to save the life of a cow where no better appliances could be had.

However, while it is possible to stop the milk fever by any means that makes it possible to pump the udder full of air, there is great danger of introducing infection at the same time that will cause inflammation and possibly result in the loss of the cow's udder.

Use of the Apparatus. — In using the milk fever apparatus the operator should first thoroughly clean his hands, likewise the cow's udder and teats, with warm water and soap, followed by a 5 per cent solution of carbolic acid or creolin. That portion of the apparatus which holds the cotton, the rubber tube leading to the milk tube, likewise the latter, must be clean, and preferably boiled 15 minutes before using, then disinfected by the use of the carbolic acid or creolin. The receptacle for holding the cotton is filled with ordinary cotton, or, better still, absorbent cotton, which may be purchased from most drug stores. The milk tube is then inserted into one of the teat openings without drawing what milk is contained, and air is pumped through the cotton into the udder. This is continued until the quarter is well distended with air, when the tube is carefully withdrawn and a tape tied around the teat tight enough to prevent the escape of the air. The same treatment is applied to each quarter. The teats are allowed to remain tied. Ordinarily within two or three hours the cow will regain consciousness and be able to stand on her feet. If the air is absorbed or escapes, so the udder is not tightly distended, the tape should be removed and another injection of air made as before. Usually two injections are all that are required. The udder should remain full of air 24 hours at least, and longer if any sign of the trouble remains. The calf of course is not allowed to suck during this time. If inflammation of the udder follows, it shows sufficient care was not taken in disinfecting the apparatus used.

CHAPTER XIX

WATER AND SALT REQUIREMENTS

Water for Cows. — Cows that are producing milk require a much larger quantity of water than is necessary for growing animals or maintenance animals. This is caused by the necessity of providing water to be used in the milk itself and for the digestion and assimilation of a larger quantity of feed, much of which is roughness. The following table shows the average amount of water consumed per day for ten days by two Jersey cows three months in milk, and by the same animals when dry and on maintenance.

WATER CONSUMED BY COWS IN MILK AND ON MAINTENANCE

Cow No.	In Milk				On Maintenance	
	Lb. Milk per Day	Lb. Fat per Day	Lb. Water per Day	Lb. Dry Matter fed per Day	Lb. Water per Day	Lb. Dry Matter fed per Day
27	26.8	1.39	77.3	28.3	14.7	9.7
62	13.3	.69	40.3	18.2	11.6	8.6

The first half of the table gives the production of milk and fat per day, the pounds of water consumed, and the amount of dry matter in the feed. The second part gives the data regarding feed and water when the two cows were dry and farrow. The cow producing 26.8 pounds of milk per day used

77.3 pounds of water, while the one producing one half that amount of milk consumed 40.3 pounds of water. The consumption of water was in practically the same ratio as the production of milk. The ration consisted of grain, alfalfa hay, and corn silage, fed in the ratio of 1 : 1 : 4, and was identical for the two animals, except it varied in quantity to meet their individual needs. The temperature of the barn and of the water used varied from 45° to 55° F.

The table below shows the enormous quantity of water consumed by a cow producing over 100 pounds of milk per day. This animal was fed a ration of alfalfa hay and grain with a small feed of silage. The water was given three times daily at a temperature of about 80° F.

WATER CONSUMED BY MISSOURI CHIEF JOSEPHINE

Days after Calving	Lb. Milk produced	Lb. Water drunk	Ration			
			Alfalfa Hay	Silage	Grain	Lb. Dry Matter fed
28	102.7	230	20	7	16	34.19
29	97.1	225	18	10	16	32.98
30	106.5	216	17	10	14	30.26
31	106.1	270	18	10	20	36.58
32	103.3	307	18	10	20	36.58
33	103.8	244	18	10	20	36.58
34	100.8	260	18	10	20	36.58

This cow produced approximately 4 pounds of milk to 1 by No. 27 as given in the preceding table, and the consumption of water was in about the same ratio. When cow No. 27 was producing milk, the consumption of water was 77.3 pounds per day, while on maintenance she consumed only 14.7 pounds.

The ration given while on maintenance was exactly the same as fed when in milk, except in regard to quantity. The large water requirement by the cow in milk, as shown in these tables, suggests, as has been found by practical experience, that it is exceedingly important to supply an abundance of good water to cows producing milk. It is evidently much more important that an abundance of water close at hand, and not too cold, be supplied to heavy milking cows than is the case when the animals are on maintenance only. Cows that are not producing milk do not need to be watered more than once a day in the winter time, and at this season they do not seem to care for it oftener than this. In the summer the consumption of water by cattle on maintenance is greater on account of the greater evaporation from the skin; and while cattle will thrive when watered once per day, they relish it oftener and will do better if supplied twice per day. Cows on heavy feed, producing large quantities of milk, should always have access to good water at least twice per day at all seasons. For the best results with dairy cows, water of good quality should be supplied close at hand, since if they be required to walk a long distance in cold weather, on account of the discomforts of exposure, they will not drink a sufficient amount to supply the demands of the body and will give a less amount of milk than they otherwise would, on account of not having consumed a sufficient quantity of water. In other words, the cow may suffer for lack of suitable water just as easily as for lack of feed.

The best source of supply for drinking purposes for dairy cattle is deep well water pumped into a tank or a cement trough. Next to this is running streams or springs. The use of ponds as a means of supplying water is not objectionable if they are so placed that there is no drainage from barnyards or about dwellings, and if the animals are not allowed to wade into the water. Ponds which are filled with contaminated drainage water or in which stock of any kind are allowed to wade and to pollute with their own excrement, are entirely unsuitable as a source of water supply. Not only should the use of such water be avoided for sanitary reasons, but the amount of water consumed under these conditions is liable to be below the real requirement of the body.

In climates where the temperature remains below freezing for long periods in the winter season, it is profitable to use some means of warming the water. It is cheaper to warm water with a tank heater constructed for the purpose by burning coal or wood than it is to supply the same amount of heat by allowing the animal to burn high-priced feed in its body. Counting the fuel value of a pound of corn, according to Armsby, at 1308 calories, one pound of corn would contain sufficient heat, if it be possible for the body to utilize all it contains, to warm 75 pounds of water from freezing to body temperature. A cow producing about 25 pounds of milk per day would require one pound of corn per day to warm the water she consumes, if it be given her at freezing temperature. Larger producers would require a correspondingly larger amount. However, there are even more important reasons for supplying warm water to heavy-producing cows than that of economy in feed. As already stated, the heavy-milking cow will not consume a sufficient quantity of cold water to make possible the maximum production of milk. When a cow takes 30 to 40 pounds of ice water into her stomach, it chills her so thoroughly that the functions of digestion and milk secretion seem to stop almost completely for a while.

If this amount or more of cold water is consumed twice daily, as would be necessary with a heavy milker, a considerable proportion of her milk secretion is necessarily lost. It is always economy to warm the water to a temperature of at least 60° when drunk by the animal.

King[1] at the Wisconsin Experiment Station found that cows given water at 70° F. drank 10 pounds more each per day than those having ice water. The actual consumption of food per pound of milk was also a trifle less on the warm than on the cold water. He shows that the heat in one pound of beef fat would be required every 3.8 days to warm the water to body temperature when the cold water is used, while the same amount of fat would heat the warmed water (70° F.) drunk, to body temperature for 7.4 days. While this loss of feed necessary to heat this water to body temperature is considerable and should be avoided, the other reasons given are still stronger.

There is some difference of opinion in regard to the advisability of having arrangement for watering cattle in the barn. In climates where the temperature is such in the winter that the animals are confined most of the time, it is a great convenience to have an arrangement whereby the animals may be watered in their stalls during the days when weather conditions are such that it is not desirable to turn them outside. A number of devices have been put on the market for this purpose. The main point to be considered in selecting such a device is the possibility of keeping the water free from contamination. If the arrangement is such that the water stands exposed to the air in the stable all the time, it soon becomes foul from absorption from the air and more especially

[1] King, Wis. Exp. Station, Bulletin No. 21.

from the cow dropping particles of food into the basin. If some arrangement is provided whereby the basins can be filled at intervals when the animals need water, and then drained out, this objection may be largely overcome. One of the most satisfactory provisions for watering cows in the stable is the use of a continuous cement manger. By this means the objections above mentioned are overcome. The manger is swept free of feed when it is desired to water the animals, and the water remaining after the animals have drunk is drawn out through a drain arranged for the purpose.

Salt Requirements. — All animals that consume large quantities of vegetable food require salt. Carnivorous animals do not have this craving, neither do human beings that live mostly upon meat. According to Bunge [1] the cause of this salt requirement by herbivorous animals is the large quantity of potassium which they consume with the plant food. The potassium is excreted through the kidneys, but while in the body a reaction takes place between the potassium and the sodium chloride or common salt, and the resulting compounds are excreted from the body. This leaves the body short in the amount of sodium chloride needed, and results in the well-known craving for common salt, or sodium chloride. Common salt is needed, according to this view, to help expel the excess of potash taken in with the vegetable food.

Babcock's Investigations. — Babcock [2] has reported extensive investigations on this subject. He kept cows in milk without salt for periods up to one year. The composition and quantity of the milk produced was not affected by withholding salt for short periods. Cows without salt showed a strong

---

[1] Bunge, *Lectures on Physiological and Pathological Chemistry.*
[2] Babcock, 22d Annual Report Wisconsin Experiment Station, p. 123.

craving for it after two or three weeks, then quieted down and gradually changed into a condition of low vitality, rough coat, and emaciated condition, which resulted finally in a complete breakdown. In most cases the cows recovered their normal condition when given an abundance of salt. He found that cows consume about one ounce per day when allowed to eat as their appetite demands. He concluded that $\frac{3}{4}$ ounce per day per 1000 pounds live weight is sufficient, with $\frac{6}{10}$ ounce in addition for each 20 pounds of milk produced. Babcock's investigations indicate that the chlorine is the essential element supplied by salt.

Feeding of Salt. — Practical observations and scientific investigations agree that salt is an essential part of the ration for the dairy cow. Salt may be supplied by mixing the proper amount regularly with the feed, or it may be placed where the animal can consume such amount as the appetite demands. From one to three ounces per day is needed, depending upon the amount of milk produced. Some feeders prefer rock salt, in which case a lump is placed where it can be licked by the animals as their appetite calls for it. The common plan of salting the cattle only at intervals of one or two weeks is not to be recommended.

CHAPTER XX

THE SOILING SYSTEM

Soiling means growing green forage, cutting it, and bringing it to the stock, in place of allowing them to eat the green feed where it grows. This system is one that as a rule goes with intensive farming and high-priced lands. It is also practiced in some other regions, on account of the conditions being such that good pasture cannot be had. In many parts of the world, for example, the greater part of Germany, cattle are never grazed. The system is also followed to some extent in the Eastern States.

When green feeds are used only to supplement pastures for part of the season, it is spoken of as partial soiling. When the animals are sustained entirely on green feed, cut and hauled to them, the plan is called complete soiling. It is not implied in either case that the animals do not receive any feed, such as grain, other than the green feed given, but that green feed constitutes the main part of the ration. On account of the comparative cheapness of land over the greater part of the United States, the system is not as yet practiced only in isolated localities. Its advantages are: —

1. Saving of land.
2. Saving of fencing.
3. Better use of the manure.
4. Animals are kept in better condition.

Saving of Land. — The greatest advantage that can be urged for soiling is the much greater returns in the way of feed that it is possible to secure from a given area. This saving of land comes about in three ways especially. The most important is that the crops are allowed to become more mature before being used. When pasture grass is eaten in an immature form, it is not given opportunity to utilize its leaves and roots to the best advantage to build up the plant. It is a demonstrated fact, for example, that the corn plant gathers the greater part of its nutrients after the plant is fully grown. If the growth should be cut off as soon as it reached the height of a few inches, the yield of feed per acre would be very small. The same is true, to a less degree, however, with grasses. Forage crops used for soiling are cut when nearly mature, but before they have become woody and unpalatable.

The second reason why pasturing does not yield as much feed as soiling crops is that in the former the plants are injured by the treading of the feet and the soiling of the grass with manure.

A third reason to be taken into account is the injury done to the land by the trampling of the stock, especially during wet weather.

By following the soiling system, Detrick,[1] whose remarkable results have been widely quoted, was able to raise all the roughness needed for thirty head of stock, of which seventeen were cows in milk, on seventeen acres. The land on which these results were obtained was in the beginning so run down that it would hardly support three animals. The fertility was built up by barnyard manure until two crops per year, together equal to about 6.7 tons of hay per acre, were produced.

[1] Farmers' Bulletin No. 242, U. S. Dept. of Agric.

At the Wisconsin Experiment Station three cows were maintained on 1.5 acres of soiling crops. Three other cows on pasture required 3.7 acres. The conclusion was drawn from this work that one acre of soiling crops equals 2.5 acres of pasture.

The Connecticut Experiment Station maintained four cows from June 1 to November 1 on 2.5 acres of soiling crops. The Kansas Experiment Station [1] reports the following results in a year of more than average good pastures. Ten cows were used.

Crop	Days Fed	Yield per Acre Lb.	Value per Acre
Alfalfa	14	77,145	$25.26
Oats	9	12,325	6.81
Corn	31	38,695	22.79
Sorghum	15	22,370	15.60
Kaffir corn	14	17,550	13.83

The average consumption was 116 pounds per head daily, and .71 acre fed one cow 144 days. The financial returns are based upon the creamery price for butter fat. On the same basis the pasture yielded $4.23 per acre.

Taking all the data reported into account, it seems conservative to say that when following the soiling system one acre will produce at least twice as much, and often three times as much food as an acre of pasture.

The second advantage of the soiling system is the saving of fencing. Fences are an item of large expense on any stock farm, especially when it is necessary to divide the land into

[1] Kansas Exp. Station, Bulletin No. 119.

small areas in order to utilize it to the best advantage. The initial cost of fences is considerable, and in addition they require almost continual repairs. In addition, the land near the fence cannot be utilized, and it is only by additional labor that unsightly weeds may be prevented from growing in this strip. This expense for fences is largely eliminated by the soiling system. In Germany, where the soiling system is almost universal, no fences at all are in use. Practically no fences are used in Denmark, where the cows are not turned loose on pasture, but are tethered out constantly.

The third advantage is the better saving of the manure. The value of barnyard manure to the fertility of the farm is recognized in every successful farming region where anything like a permanent system of agriculture has been established. The actual value of the excreta passed by a cow in a year for fertilizing purposes is about $30, if all is preserved. If the animals are on pasture, nearly half of this will be dropped in the pasture. In most cases a much greater value would be realized from applying this fertilizer to other parts of the farm more in need of improvement. When the cattle are kept housed, as is usually done in soiling, the manure may be preserved and applied when most needed. On some farms this is a point of great importance.

Animals are kept in better condition, as a rule, where the soiling system is followed than when they are pastured. This results from a more regular feed supply, ample at all times to allow of profitable production. The serious decline in milk production, for example, experienced in midsummer by most cow owners following the pasturing plan, is almost entirely avoided. At the same time, the labor and expense of getting these additional results must be taken into consideration.

Objections to the Soiling System. — There are two objections to the soiling system which prevent its wider adoption.

(1) The labor problem.

(2) Difficulty of providing a suitable series of crops and adjusting the amounts of each.

The Labor Problem. — First of all the soiling system involves a much greater expense for labor than when the pasturing system is used. The green feed must be cut each day, or at the outside every second day. The common practice is to cut it daily, except Sunday, when the animals are fed from a larger amount prepared the previous day. Since the amount required per day is around 100 pounds per cow, the weight to be handled is considerable, making the labor heavy.

A further difficulty about the labor feature is the regularity required of attendants, which is difficult to secure. In addition to preparing the green food, the labor of handling the manure and caring for the animals from day to day must be considered. The difficulty of securing a suitable succession of crops in about the right quantity requires careful planning when complete soiling is practiced. There must be a succession ready to use all the time, and in about the proper quantities. A surplus may usually be made into hay, but this is not always convenient.

Crops for Soiling. — The crops to be grown for soiling purposes depend naturally upon the local conditions, and no general statement can be made to cover all conditions. Information regarding the best crops for the purpose can be had from the nearest experiment station. Shaw[1] gives the following list of suitable crops.

For Canada and the Eastern States north of the Ohio River,

---

[1] *Soiling Crops and the Silo.* Orange Judd Co.

winter rye, alfalfa, medium clover, mammoth clover, peas and oats, corn, sorghum, millet, and field roots.

For the northern part of the Mississippi valley the same, except alfalfa is omitted.

For the Southern States winter rye, winter oats, crimson clover, corn, sorghum, cowpeas, and rape are suggested. In sections where alfalfa can be grown it should be included by all means. In the southwestern part of the United States more attention should be given to alfalfa, soy beans, and Kaffir corn. The Pennsylvania Experiment Station [1] recommends the following succession of crops and acreage for ten cows where alfalfa can be grown.

Crop	Area	When to be fed
Rye	½ acre	May 15–June 12
Alfalfa	2 acres	June 1–June 12
Clover and timothy	¾ acre	June 12–June 24
Peas and oats	1 acre	June 24–July 15
Alfalfa (2d crop)	2 acres	July 15–Aug. 1
Sorghum and cowpeas (after rye)	½ acre	Aug. 11–Aug. 28
Cowpeas (after peas and oats)	1 acre	Aug. 28–Sept. 30

[1] Bulletin No. 75.

CHAPTER XXI

FEEDING FOR MILK PRODUCTION [1]

There are two factors which largely control the economical production of milk. One is the adaptability of the cow used for this purpose, and depends upon her individual and breed characteristics. The other is the amount and kind of food eaten. The problem confronting the dairyman is the production of the largest amount of milk and butter at the least expense. In order that this may be realized, both the important factors mentioned must receive careful attention.

In most cases the largest direct expense is for feed. Every one familiar with the prevailing conditions knows a large amount of feed is used without producing the returns it should. It would be safe to say that the average yearly milk production per cow could be increased one half or three fourths by following better methods of feeding.

Turning on Pasture in the Spring. — Every owner of a cow welcomes the time when the animal can be turned out to pasture. Not only is the labor and expense connected with winter feeding done away with, but each cow is expected to give the best results of the year on grass. In changing from

[1] The author does not attempt to give more than a brief outline of this subject, and that mostly from the standpoint of practical feeding. The reader is referred to *Feeds and Feeding*, by Henry; *The Feeding of Farm Animals*, by Jordan; *The Principles of Animal Nutrition*, by Armsby; and the *Ernährung der Landwirtschaftlichen Nutzthiere*, by Kellner, for more extended discussions of the subject.

dry feed to grass, it is best to go somewhat slowly, especially with heavy-milking cows. The young, immature grass, such as we have in early spring, contains a large amount of water and a small amount of dry matter, and it is almost impossible for a heavy-milking cow to eat enough of such feed to supply the necessary amount of nutrients. Wheat and rye pastures are of the same nature. Another reason for putting cattle on pasture gradually rather than suddenly is the effect on the taste of milk. When a cow is changed at once from a grain ration to grass, a very marked taste is developed in the milk, while if this change in feed is made gradually and not suddenly, the change in the taste of the milk is scarcely noticed.

Grain Feeding while on Pasture. — There is some difference of opinion on this question from the standpoint of economy. There is no question but that a cow will produce more milk if fed grain while on pasture, and if a large yield is of more importance than economy of production, grain should certainly be fed. The cow that gives a small average quantity of milk will produce but little more, if fed grain while on pasture. However, with the heavy-producing cow the case is quite different, and it is necessary that she be fed grain, or she will not continue on the high level of production long. The necessity of feeding grain to the high-producing cow arises from the fact that she cannot secure a sufficient amount of nutrients from the grass alone, and must have some concentrated feed in the form of grain in order to continue to produce large quantities of milk.

Experiments made by the Cornell Experiment Station, covering four years, showed that, while an increase in milk yield was secured from grain feeding, it was not economical to produce it in this way. In their tests only about an additional pound of milk was secured for each pound of grain fed. In these experiments the pastures produced an abundance of nutritious grasses. It was observed, however, that the cows fed grain during the summer gave better results after the grazing period was over than those not having received grain. This is also a matter of common observation, and should be taken into account in considering the advisability of feeding grain. The point is that the cows fed grain stored a considerable quantity of surplus nutrients on their body, which were afterwards available for the production of milk. A Jersey cow that is giving as much as 20 pounds or 10 quarts per day, or a Holstein or Shorthorn giving 25 pounds or more daily, should be given some grain. The practice of the author in regard to feeding on pasture is about as follows: —

Jersey or Guernsey cow producing: —

20 lb. milk daily	3 lb. grain
25 lb. milk daily	4 lb. grain
30 lb. milk daily	5¼ lb. grain
35 lb. milk daily	7 lb. grain
40 lb. milk daily	8 lb. grain

Holstein, Shorthorn, or Ayrshire producing: —

25 lb. milk daily	3 lb. grain
30 lb. milk daily	4 lb. grain
35 lb. milk daily	5¼ lb. grain
40 lb. milk daily	7 lb. grain
50 lb. milk daily	9 lb. grain

It must be kept in mind that this applies only when pastures are abundant. Where a small amount of grain is fed to a cow on pasture, corn is as well adapted as anything else where it is cheaper than other feeds, since, on account of the comparative narrow nutritive value of the grass, the corn does not unbalance the ration. However, in case of feeding large quantities of grain, for example, 5 pounds per day, or above, other feeds, containing more protein, should be used in part, such as: bran, gluten meal, oats, or cottonseed meal.

Summer Conditions to be Maintained as near as Possible throughout the Year. — Soon after the cows are on pasture, usually in the latter part of May or the first part of June, they usually reach the maximum production of milk for the year. This suggests that what the dairyman must do in order that the production of milk may be the largest is to imitate these early summer conditions as far as possible throughout the remainder of the year. This is what the careful dairyman and skilled feeder does, and the results correspond closely to the success with which these summer conditions are maintained. The summer conditions which bring about the maximum production, and which are to be maintained as far as possible through the year, are described in the following statements: —

1. An abundance of palatable food.
2. A balanced ration.
3. Succulent feed.
4. Moderate temperature.
5. Comfortable surroundings.

Providing for Periods of Short Pasture. — As long as fresh pasture grasses are abundant, the ordinary cow is about as well provided for as she can be to produce milk economically. Unfortunately the season of abundant pasturage is often short. In many localities a dry period, often of several weeks, occurs during the middle or latter part of the summer, and the pastures become short and insufficient to maintain a full flow of milk. This season is often the critical time of the year for the dairy cow. It is probable that as much loss occurs one year with another by lack of feed at this time as occurs from improper feeding during the winter season. When the season of dry feeding arrives, the farmer expects to feed his stock, and is prepared for it. On the other hand, as long as the cattle are on pasture and the field work is pressing, the tendency is to let the cows get along the best way they can.

On a large proportion of the farms, the cows are fresh in the spring, give a good flow of milk while the pastures are good, but when hot weather and short pastures come, the flow drops one half or two thirds, and the cows are producing but a small amount in the winter when the price is the highest. It is almost impossible to restore the flow of milk to the original amount after it is once allowed to run down from lack of feed. To make large returns from the cow a large yearly production must be had; and to do this, the flow of milk must be kept up ten or eleven months in the year.

It is possible to hold up the milk flow by heavy grain feeding, but this is unnecessarily expensive. Provision should always be made to have green crops on hand that may be cut and fed when needed, or to have silage available. It is the nature of blue grass to grow freely in early summer, then to rest until fall. This leaves a period in the summer where blue grass is depended upon for pasture from about the middle of July to the middle of September, when the pasture is apt to be short.

Corn is in many ways the best crop for summer soiling. The main difficulty is, it does not come on early enough. Even the early varieties are hardly mature enough to feed before August 1, and something is often needed earlier. Sweet corn is a good feed, but does not yield heavily. Second-growth clover, millet, or alfalfa can be used if available. After August 1, in the corn belt, corn and sorghum are the best crops for supplementing pastures.

Sorghum yields immense crops, and if a surplus is on hand it may be made into hay profitably. A yield of from 15 to 25 tons of green sorghum per acre is not unusual on good land.

Green crops fed as a supplement to pasture may be fed in the pasture or in the barn lot, but as a rule are fed most economically in the barn. The cows remain inside long enough at milking time to eat their portions.

As a rule the most economical method of supplying feed to help out the short pastures of midsummer and fall is to feed corn silage. Silage will keep in good condition for summer feeding, with no loss except on the surface. If it is not needed during the summer, it may be covered with the new silage and kept until wanted. Corn furnishes a larger yield of dry matter per acre than any crop that can be ordinarily grown for summer feeding, and has the further advantage of being on hand as early as wanted when in the form of silage.

It is handled more economically also than soiling crops, since it is cut all at once, and not every day, as is necessary with soiling crops.

The great problem in winter feeding, as already stated, is, in general, to maintain the conditions of early summer. It is entirely feasible to maintain practically these summer conditions throughout the entire winter on any farm when the subject is properly understood and the necessary arrangements made.

Amount of Feed. — The first condition given as typical of the summer feeding is an abundance of palatable food, and on this point is made one of the most common mistakes in feeding cows. In producing milk, the cow may be looked upon in a way as a milk-producing machine which we supply with a certain amount of raw material in the form of feed, and this raw material is manufactured into milk. The same rule holds as in the running of any other manufacturing plant; it is run most economically near its full capacity. Every one who feeds animals should thoroughly comprehend that, first of all, the animal must use a certain proportion of its food to maintain the body. This is the first requirement of the animal, and it is the first use to which it puts its food. This is called the ration of maintenance, and it is practically a fixed charge. That is, it is practically the same whether the animal is being utilized for maximum production, or is merely kept without producing any milk at all.

In the case of an ordinary dairy cow this ration of maintenance amounts to from 50 to 60 per cent of all she can consume. In the case of a heavier producing animal, for example, one producing from 1 to $1\frac{3}{4}$ pounds of butter fat per day, this ration of maintenance amounts to from 40 to 50 per cent of the total feed of the animal. It should be clear that, after going to the expense of giving the animal the necessary amount to keep her alive, it is the poorest economy to refuse to furnish the other 40 to 60 per cent, which she would utilize exclusively for milk production. On the average farm this is one of the most common mistakes made. The importance of liberal feeding for economical production can be easily understood from the following illustration.

The first illustrates the proper feeding of a heavy-producing cow, which is the one usually underfed. The line a to c represents the total capacity of the animal for food, or a full ration. The first half, from a to b, represents the amount of food required to maintain the animal's body, or the ration of maintenance. The second half, that portion from b to c, represents the proportion of the food used for the production of milk. In this case there is no fat being produced on the animal's body, and the cow is supposed to be of such dairy quality that all the feed she can eat in excess of that required for maintenance is used for milk production.

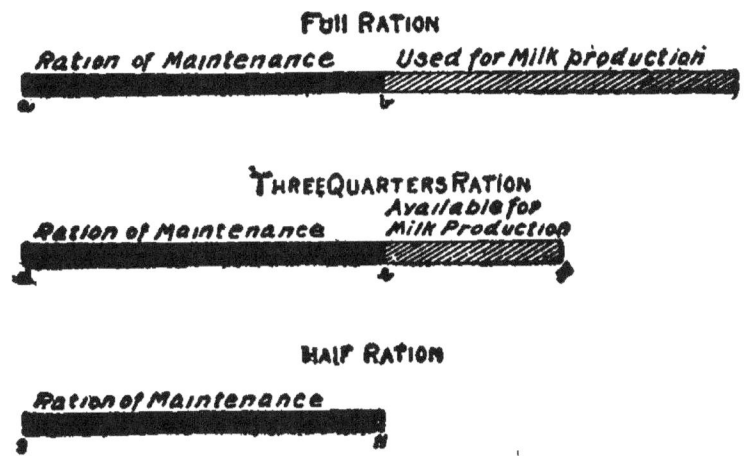

The line below represents what would happen if the feed of this animal is reduced one fourth. The ration of maintenance remains practically the same as in the first case. The amount represented by the line d to e is the amount required to maintain the animal's body, which is the same quantity as in the first case; however, the cut of one fourth in the ration will be seen to come entirely on that available for milk production, and reduces that amount one half.

Suppose that the ration of such a cow be still further reduced to one half of the full ration, or that required for maintenance alone, as represented by the third line. In this case the cutting down of the ration one half would remove all available feed for milk production. However, the animal would not cease producing milk at once. This is a point of great importance in feeding cows, and a lack of such knowledge leads to serious errors in feeding. The milk-producing function is so strong that the cow will continue to produce milk for some time, even when the feed is insufficient, utilizing the reserve material which has been accumulated in the body in the past. This always happens in the case of a heavy-milking cow during the first few weeks after the birth of the calf. At this time, it is not generally possible, and not desirable on account of the condition of the animal, to feed her a sufficient quantity of feed to supply the nutrients necessary to produce the milk; and even if the feed was offered, the appetite is not usually strong enough to cause the necessary amount of feed to be taken to prevent loss in weight. As a rule, all heavy-milking cows decline in weight for the first two or three weeks, and occasionally for ten weeks, after calving, which means that milk production has been in excess of the feed supplied for that purpose. The same thing happens in the case of the cow that is not fed a sufficient ration for the amount of milk that she is producing. She may continue to produce considerable milk for a while by drawing on the reserve material of the body, but as soon as this is exhausted, the production of milk must come down to the amount available for this purpose, above the ration of maintenance. When the feed is in excess, the cow begins to store reserve material on her body. If the amount of milk produced by a cow varied directly with the feed, and she did not store up nutrients at one time and draw on reserve material at another, it would simplify the problem of feeding very much, and result in more economical feeding at all times.

How to avoid Overfeeding. — While the statement and illustration given apply to one class of dairy cows, there is another class to which it does not apply, and with which it would lead to a serious mistake in feeding from an economical standpoint. This group includes those of lower productive capacity, which are liable to be overfed, especially when they are in the herds of dairymen, who realize the necessity of liberal feeding. The proper feeding of this group of animals can be made clearer by the following illustration: —

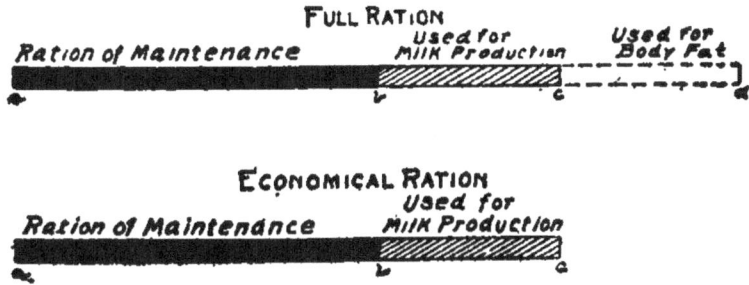

The line *a* to *d* represents the amount of feed that an animal of this class will consume; *a* to *b* represents the ration of maintenance, as before. In this case, however, the capacity for making milk is not equal to the capacity of the animal for utilizing feed in excess of that required to maintain the body. The amount which the animal is capable of utilizing for milk production is represented by that portion of the line *b* to *c*, while the animal's appetite is equal to the total line *ad*. This gives a surplus, *c* to *d*, which is not utilized for milk production, but which will be used for storing fat on the animal's body; and we will have the cow gaining in weight while she is producing milk. This gain in weight will be of no service as far as milk production is concerned, except that it is of some value as a reserve material to be drawn upon at some other time when feed is not supplied in sufficient amounts; but it is not economical nor desirable to fatten dairy animals with the expensive feeds which are fed cows in milk. That portion of the feed represented by the line *cd* should be taken from the ration. This means reducing her feed to take off the amount used for storing fat on the body; in other words, to feed her only what she will utilize for milk production. This means feeding enough to maintain a practically uniform body weight. In every large herd where the amount fed is not carefully regulated, we find errors made in both these classes. We find the heavy-producing cows being underfed, and we find the light-producing cows being overfed and allowed to accumulate fat.

Relation of Live Weight to Proper Feeding. — The live weight of a cow is one good index of whether the cow is being fed a proper amount or not; but good judgment must be used in regulating the ration by observing this condition. We must expect that a cow will lose weight in the first few weeks of her milking period; but after this period is past, there is no reason why she need to change much in weight for several months, and this is the period when the greater part of the milk production is secured. It will not mean, of course, that the animal should not be allowed to gain in weight during the latter end of the milking period. This is necessary on account of the development of the fetus, and since it is natural for the animal to carry some fat on her body at calving time.

It does mean, however, that in order to feed a herd of cows economically it will not to do feed them all the same quantity of grain whether they are giving a gallon of milk a day or whether they are giving four gallons, and it means that when a cow in the middle of her lactation period is putting on weight she is being fed more than she needs and will give just as much milk if the feed is cut down somewhat. It also means that if a certain animal is losing in weight sufficient feed is not being given, and if the deficiency is not supplied it will not be long before the milk production will come down to correspond with the amount of feed available.

Feeding as Individuals. — In connection with the discussion of the amount to feed, it needs to be pointed out that it is only possible to feed a bunch of cows economically when they are fed as individuals, and not as a herd. A too common practice, even in otherwise well-managed herds, is for all animals to be fed the same amount of grain, regardless of the time they have been in milk or the quantity of milk individual cows are producing. Such feeding always lacks economy, as the high-producing cow does not get enough, and while she may milk very well for a short time, she soon comes down to a lower level, while the lighter producing cow usually gets too much and accumulates fat. The production of many herds could be increased to a large extent without any increase in the amount of feed used by distributing it properly among the individuals. It requires some attention to adjust the ration to each individual, but the time spent in this direction gives good returns.

Amount of Grain and Roughness to Feed. — The cow, being adapted by nature for consuming bulky feeds, does not feel satisfied unless there is sufficient bulk to the ration given at all times. An animal that is fed too much grain in proportion to the amount of roughness may seem hungry, while she really has a sufficient amount of nutrients, but so concentrated that it does not have sufficient bulk. The cow should be fed practically all the roughness she will eat up clean, at all times, and the difference in rations fed to cows producing different quantities of milk should be in the grain ration.

A cow on a good ration of roughness will maintain herself and produce a certain amount of milk. If she be a cow of much dairy capacity, she will not produce milk to anything like her maximum without having a portion of her ration in the form of concentrates. The point is, the milk-producing function has been developed to such an extent that it is impossible for the digestive apparatus of the cow, efficient as it is, to extract sufficient nutrients from coarse feeds to supply the enormous drain upon the body resulting from the secretion of large quantities of milk.

The mistake is at times made of assuming that cows all receive the same ration when a uniform grain ration is fed. The difference in the amount of roughness consumed is generally overlooked in these cases, since the animals can eat at will. If a grain ration be increased which is already ample the animal consumes less roughness and may not consume any more nutrients than before, although usually such is the case on account of the greater palatability of the concentrates. In herds where all cows receive the same grain ration, close observation will show that the light milkers are consuming less forage than the heavier milk producers. Since roughness is usually a cheaper source of nutrients than grains, it is desirable to have a liberal amount of this class consumed. The amount can be regulated by giving the animal all she will consume of the roughness, and in addition concentrates to supply the nutrients necessary for the amount of milk she is producing.

The most accurate means of determining the ration needed is by calculation based upon the feeding standards. However, the following rules are of service as a general guide for practical feeding: —

1. Feed all the roughness the cows will eat up clean at all times.

2. Feed one pound of grain per day for each pound butter fat produced per week, or one pound grain daily for each three pounds of milk.

3. Feed all the cows will take without gaining in weight.

The rule regarding the amount of grain to feed per day to each cow applies only when good roughness, such as corn silage and clover, cowpeas, or alfalfa hay is used. The second part of the rule, in regard to feeding one pound of grain for three pounds of milk, will not work out in all cases. For a heavy-milking Holstein cow, for example, this gives a little too large a quantity of grain, and with a cow giving very rich milk it is a little too low. It applies best to cows producing milk of about average composition.

The rule based upon the butter fat produced per week is the best, as it applies to any breed. If the roughness be timothy hay or corn fodder, considerable more grain must be fed in proportion to the amount of milk produced.

Home-Grown Balanced Rations. — One reason why the average farmer makes a mistake of feeding his cows rations that are not properly balanced is that it is easier, or he thinks it is, to grow feeds that are excessively rich in carbohydrates and lacking in protein. This comes about principally by the large amount of corn and timothy hay grown and used. It is impossible from these feeds to make a ration that supplies the necessary nutrients to produce much milk.

It is possible to make a fairly good ration, using these feeds for roughness; but it is only possible to do so by feeding large quantities of mill feeds that are rich in protein. The thing for the farmer to do is to raise the feeds he requires on his own farm, as far as possible; and it is possible in many localities to produce practically all that is needed to make a balanced ration. The place to begin in considering the feeding of an animal always is with the roughness, since the character of the roughness determines to a large extent the kind of grain it is advisable to feed.

The cheapest source of protein usually is leguminous hays, including clover, alfalfa, and cowpea hay. If an abundance of any one of these hays can be grown, the problem of making an economical balanced ration is very much simplified. Corn is a staple crop in most parts of this country where many dairy cows are kept. Corn silage, a legume hay, and corn for grain make a ration that can be grown in most localities. These alone make a satisfactory ration for ordinary producing cows. The addition of a small quantity of a concentrate rich in protein, like bran or cottonseed meal, makes a ration adapted to heavier producers of milk.

Succulent Feeds. — Another reason why cows do well on good pasture is the character of the ration given. Green feeds have that property called succulence. Such feed has a value outside of the actual nutrients it contains, on account of the favorable effect upon the digestion of the animal. In the corn belt corn silage furnishes the best means of supplying this class of food. In some sections, especially north of the corn belt, the growing of root crops is widely practiced, and supplies this desirable addition to the ration in an entirely satisfactory form.

The Balanced Ration. — A ration is said to be balanced when the two classes of nutrients are present in the right proportions. The cow produces milk abundantly when on fresh pasture grasses. One reason for this is that growing grasses constitute a balanced ration. The winter ration is liable to have the nutrients out of proportion. In the corn belt the lack of protein is the most common deficiency in the ration, brought about by the large use made of corn and corn fodder. The wide use of timothy hay is also responsible for many rations lacking in protein. A cow secreting milk must use a certain amount of protein to form the curd or nitrogenous part of the milk. No other material can take the place of protein for this purpose. A careful study of the composition of feeds and the method of computing rations aids the feeder to prepare the ration to the best advantage.

Milk Secretion due to Stimulation. — The cow is too often fed with the idea in mind that the production of milk is directly dependent upon the food consumed and that the more feed that can be gotten into her, the more milk there will be produced. According to this view, the main question is one of providing the proper amounts of suitable feeds. The correct way to look upon the cow is from an entirely different point of view. According to the view of the author, as already expressed, the production of milk is dependent upon some stimulating principle that is formed in the body and which acts upon the udder gland. A good dairy cow is one that has the stimulation to produce large quantities of milk. Any animal after violent exercise, for example, the dog after a hard day's hunting, is exceedingly hungry, and consumes a correspondingly large amount of food. The dairy cow is a parallel case. A high-class dairy cow produces a large amount of milk on account of having a strong stimulation applied to her udder. This removes a large amount of nutrients from the body; as a result, like the animal that has had violent exercise, she has a strong appetite and consumes a large quantity of food. This capacity is inherited, and cannot be put into a cow by feeding. The keen appetite and strong digestion of the good dairy cow do not cause the heavy milk production, but are a result of it. The feeder's business is to make the best use of what capacity for producing milk the cow inherits, and supply all the nutrients necessary to use this milk-making capacity to the limit. If the animal is not given sufficient feed, the reserve material is drawn upon for a certain length of time, after which the stimulation to produce milk gradually declines. After the milk flow has declined, it is usually found impossible to bring it back to the former higher level.

Condition of the Cow at Calving. — One of the most important factors in obtaining a large production of milk is to have the animal in a good condition when fresh. The fat and other reserve material that are accumulated in the body act as a reserve, which is drawn upon during the first few weeks especially. Liberal feeding before and good condition at calving also start the animal at a higher level of milk production than is the case when the animal is not in proper condition, and mean larger possibilities for production throughout the year. When the cow is in moderate flesh, or a little more than moderate, and has had from four to six weeks' rest before calving, the amount of milk she will be giving at the end of two or three weeks after calving is dependent only in part upon the manner of feeding at that time. It is far more under the control of the natural stimulation of the animal, and the problem for the feeder is to first get the animal in this condition, observe what she produces at her maximum, then adjust the ration to the amount produced and maintain the milk flow at a high level as long as possible. However, if she does decline in milk when receiving a sufficient ration, it is useless to further increase the feed and expect to restore the milk flow. On the contrary, if it is certain that the ration has been ample or more, it is economy to reduce the ration in about the same proportion.

Some Suitable Rations. — The following rations are suggestions for the beginner rather than for the expert dairyman. They supply the necessary material to produce milk economically. If the cow will not give a good flow of milk in the early part of the milking period and when fed a liberal amount of one of these rations, it indicates that she is not adapted by nature to be used as a dairy cow and should be disposed of. The amounts given are about right for the cow giving from 20 to 25 pounds of average milk per day. For heavy-milking cows these rations would have to be increased, especially in the grain, and for light-milking cows the grain should be decreased. In making up these rations it is designed that the cow be given all the roughness she will eat, and a sufficient amount of grain to furnish all the proper amount of digestible material. It is not designed that these rations should be sufficient or the best adapted for cows of unusual dairy capacity, and certainly not for cows that are being fed for making records where a maximum production is desired.

The figures given are per day. It is expected that the grain ration will be mixed in quantities and the animals fed from the mixture.

SOME GOOD DAIRY RATIONS

Ration 1

	Lb.
Corn silage	25
Clover hay	10
Corn	4
Wheat bran	4

Ration 2

Corn silage	30
Alfalfa or cowpea hay	10
Corn	6
Wheat bran	2

Ration 3

Clover hay	20
Corn	4–5
Wheat bran or oats	2–4

Ration 4

Clover hay	20
Corn and cob meal	6
Gluten or cottonseed meal	2

Ration 5

Alfalfa or cowpea hay	10
Corn fodder	10
Corn	6
Wheat bran	2

Ration 6

Alfalfa or cowpea hay	15–20
Corn	8–10

Ration 7

	Lb.
Mangels or sugar beets	25
Corn stover	10
Clover hay	6
Corn meal	3
Wheat bran	2
Brewer's grains	2
Gluten meal	1

Ration 8

Corn silage or roots	30
Clover hay	12
Oats or wheat bran	4
Ground peas or gluten meal	3
Brewer's grains	2

CHAPTER XXII

FEEDING FOR MILK PRODUCTION (*Continued*)

THE CALCULATION OF RATIONS, FEEDING HIGH-PRODUCING COWS, DISCUSSION OF COMMON FEEDS

A dairy cow uses feed for the following purposes: —

1. For maintaining the body.
2. To supply the material for milk.
3. For development of the fetus.
4. For growth in case the animal is immature.
5. At times to produce gain in weight.

For each of these purposes three general classes of food material are required.

1. Protein or nitrogenous material.
2. Carbohydrates and fat to supply the heat and energy.
3. Ash or mineral matter.

The main problem of feeding is to supply the proper amount of the food material of the three classes in the least expensive form. It is evident that the first step is to know what the animal requires for food and how to prepare a ration that will meet this demand.

How a Chemist divides Feed. — When a chemist makes an analysis of any foodstuff, clover hay, for example, he determines the amount of water, protein, ash, crude fiber, nitrogen free extract, and fat the substance contains.

Water. — All feeds, even those apparently dry, like corn or hay, contain a portion of water varying from 10 to 15 per cent with this class. Roots, such as beets and turnips, contain around 90 per cent of water. The water in the feed eaten serves the same purpose as ordinary water consumed by the animals.

Ash. — This is the mineral part of the plant substance remaining after the material is burned. It makes up the greater part of the bones, and is a necessary part of all lean meat.

Protein. — This important constituent is known by the fact that it contains nitrogen. It serves the purpose of building up tissue in the body, such as muscle, skin, etc., and constitutes the curd of milk. Lean meat and the white of an egg are familiar examples of nearly pure protein. All feeds contain more or less protein. Among hays, clover, alfalfa, cowpea, or soy bean contain the largest amount. Among the common concentrates linseed meal, cottonseed meal, and wheat bran contain relatively large quantities. A certain amount of protein is indispensable in a ration, as nothing else can be substituted for it by the animal.

Crude Fiber. — This is the woody part of the plant, which is the least digestible. The amount of this constituent increases with the age of the plant, and is large in feeds like hays and corn stover, and small in concentrates like corn and linseed meal.

Nitrogen Free Extract. — This is a term given to a class of substances much like the crude fiber in composition, but which are much more easily digestible. The greater part of this group is composed of starch.

Fat or Ether Extract. — That part of the foodstuff that will dissolve out in ether is called ether extract. It consists mostly of fats, but sufficient other products to make it somewhat inaccurate to call this group fat, although this term is often used.

The crude fiber, nitrogen free extract, and fat all serve much the same purpose in the body. They supply heat to keep the body warm, material to be built into fat and to be burned or oxidized in the body to furnish energy. All feed stuffs contain these same constituents, but in widely varying quantities.

Digestibility. — However, the animal is not able to digest all of the substances in any foodstuff. The proportion of the protein, for example, that may be used depends largely upon the nature of the feed under consideration, the grains being more thoroughly digested than the hays. The amount of each of the substances given that can be digested from any feed stuff by the animal is determined by what are termed digestion trials. The chemist makes such a trial by analyzing the food consumed during a certain period by an animal, and at the same time collecting all the dung excreted and analyzing that to find out how much passes through the alimentary canal. The difference between the amount consumed and the amount voided is called digestible. Such tests have been made of all common feeding stuffs, so the practical feeder has data at hand regarding both the composition of feeds and their digestibility to serve as a guide in preparing suitable ration.

The Feeding Standard. — The many analyses which have been made enable us to know how much of each of the several constituents is contained in all common feeds on the average.

It is also known that the cow needs all of these constituents. The next question is as to how much of each constituent is needed to supply what the cow must have to enable her to produce the maximum amount of milk. This problem has been worked on for many years by able investigators, and a fairly accurate knowledge of the subject has resulted. A statement of the food requirements of the animal is known as a feeding standard.

The feeding standard prepared by Wolff, and known in the revised form as the Wolff-Lehmann standard, has been the most widely used. The standard of Haecker has also been widely used in this country, and is an improvement over the first mentioned, in so far that a provision for maintenance is first made, then an allowance added according to the amount of milk produced. In using these feeding standards it was assumed that a pound of digestible protein in one feed stuff is equal to that amount in every other. For example, a pound of protein in clover hay is considered equal in feeding value to the same amount of protein in oil meal. More recent work by Kellner and Armsby has shown that this is erroneous and that a pound of digestible nutrients in a concentrate like corn or oil meal is more valuable than the same amount in a coarse food like hay. This difference in value is due to the increased expenditure of energy consumed in digesting the coarser feeds. The latter authorities also consider that a portion of the protein which is in the form known as amids cannot be used by the animal. Table 10 gives the digestible protein and energy value of common feeding stuffs according to Armsby.[1]

Armsby has also prepared a tentative feeding standard for

[1] Farmers' Bulletin 346, U. S. Dept. of Agric.

TABLE 10

Dry Matter, Digestible Protein, and Energy Values per 100 Pounds

Feeding Stuff	Total dry Matter	Digestible Protein	Energy Value
Green fodder and silage:—	*Pounds*	*Pounds*	*Therms*
Alfalfa	28.2	2.50	12.45
Clover — crimson	19.1	2.19	11.30
Clover — red	29.2	2.21	16.17
Corn fodder — green	20.7	.41	12.44
Corn silage	25.6	.88	16.56
Hungarian grass	28.9	1.33	14.76
Rape	14.3	2.16	11.43
Rye	23.4	1.44	11.63
Timothy	38.4	1.04	19.08
Hay and dry coarse fodders:—			
Alfalfa hay	91.6	6.93	34.41
Clover hay — red	84.7	5.41	34.74
Corn forage, field cured	57.8	2.13	30.53
Corn stover	59.5	1.80	26.53
Cowpea hay	89.3	8.57	42.76
Hungarian hay	92.3	3.00	44.03
Oat hay	84.0	2.59	36.97
Soy bean hay	88.7	7.68	38.65
Timothy hay	86.8	2.05	33.56
Straws:—			
Oat straw	90.8	1.09	21.21
Rye straw	92.9	.63	20.87
Wheat straw	90.4	.37	16.56
Roots and tubers:—			
Carrots	11.4	.37	7.82
Mangel-wurzels	9.1	.14	4.62
Potatoes	21.1	.45	18.05
Rutabagas	11.4	.88	8.00
Turnips	9.4	.22	5.74

Feeding Stuff	Total dry Matter	Digestible Protein	Energy Value
Grains:—	*Pounds*	*Pounds*	*Therms*
Barley	89.1	8.37	80.75
Corn	89.1	6.79	88.84
Corn-and-cob meal	84.9	4.53	72.05
Oats	89.0	8.36	66.27
Pea meal	89.5	16.77	71.75
Rye	88.4	8.12	81.72
Wheat	89.5	8.90	82.63
By-products: —			
Brewers' grains — dried	92.0	19.04	60.01
Brewers' grains — wet	24.3	3.81	14.82
Buckwheat middlings	88.2	22.34	75.92
Cottonseed meal	91.8	35.15	84.20
Distillers' grains — dried:			
Principally corn	93.0	21.93	79.23
Principally rye	93.2	10.38	60.93
Gluten feed — dry	91.9	19.95	79.32
Gluten meal — Buffalo	91.8	21.56	88.80
Gluten meal — Chicago	90.5	33.09	78.49
Linseed meal — old process	90.8	27.54	78.92
Linseed meal — new process	90.1	29.26	74.67
Malt sprouts	89.8	12.36	46.33
Rye bran	88.2	11.35	56.65
Sugar-beet pulp — fresh	10.1	.63	7.77
Sugar-beet pulp — dried	93.6	6.80	60.10
Wheat bran	88.1	10.21	48.23
Wheat middlings	84.0	12.79	7.75

dairy cows, based upon the most recent investigations regarding nutrition. He bases this standard upon the amount of digestible protein (amid free) and the energy value, which represents the carbohydrates and fat together. He uses the term "therm" to represent an energy value of 1000 calories. He first estimates the protein and energy value required for maintenance, and to this adds the amount of each necessary to supply what is needed for the milk. The maintenance requirements are as follows: —

Live Weight	Digestible Protein Pounds	Energy Value Therms
500	.30	3.80
750	.40	4.95
1000	.50	6.00
1250	.60	7.00
1500	.65	7.90

The maintenance requirement naturally increases with the size of the animal, but not in direct proportion.

Armsby suggests .3 therm in energy value and .05 pound of digestible protein for each pound of milk. This standard for milk production is based upon milk containing about 4 per cent of fat and 13 per cent of total solids. However, no one figure given could be accurate for milk of all degrees of richness.

Calculating a Ration. — Let it be assumed that it is desirable to calculate the ration for a 1150-pound cow producing 30 pounds of milk per day. According to the preceding table, there would be required approximately the following for maintenance: —

  Digestible Protein . . . . . . .55 lb.
  Energy . . . . 6.50 therms

For the production of 30 pounds of average milk there would be needed: —

  Digestible Protein (30 × .05) . . 1.50 lb.
  Energy (30 × .3) . . . . . . 9.00 therms

FEEDING FOR MILK PRODUCTION

The total requirements then are as follows: —

	Digestible Protein	Energy Value
For maintenance	.55	6.50
For milk production	1.50	9.00
	2.05	15.50

The problem is to find a ration that contains this amount of digestible protein and has this energy value. Other problems also enter into the question, such as bulk and the comparative cost of several feeds available. In calculating a ration we always begin with the roughness, since this class usually furnishes the nutrients cheaper than the concentrated feeds such as grain and mill products. Further, on most farms considerable roughness is on hand that must be used to the best advantage, and as already pointed out, the cow is adapted for consuming roughness and must have a certain bulk to her ration at all times. We will assume that on the farm where the foregoing ration is to be fed corn silage, clover hay, and corn are on hand, and wheat bran and cottonseed meal may be purchased if necessary to provide the proper ration.

A good ration of roughness in quantity would be corn silage 35 pounds and clover hay 9 pounds. Using Table 10, this would give the following: —

	Digestible Protein Pounds	Energy Value
35 lb. silage	(.35 × .88) .31	(.35 × 16.56) 5.79
9 lb. clover hay	(.09 × 5.41) .48	(.09 × 34.74) 3.12
	.79	8.91

This leaves 1.26 pounds of digestible protein and 6.59 therms of energy to be supplied by the grain. Since corn is grown on the farm, we will use it as far as possible in making up the grain ration. The amounts to be used can only be found by trial. We will start with the following: Corn 6 pounds, bran 3 pounds, cottonseed meal 1 pound. This would give us the following amount of protein and energy values: —

	Digestible Protein	Energy Value
35 lb. corn silage	.31	5.79
9 lb. clover hay	.48	3.12
6 lb. corn	.40	5.33
3 lb. bran	.30	1.44
1 lb. cottonseed meal	.35	.84
	1.84	16.52
Required	2.05	15.50

This gives too much energy value and is a little low in the protein. Since cottonseed meal is the highest in protein, we will cut out the bran and increase the cottonseed meal by the addition of another pound. We would then have the following: —

	Digestible Protein	Energy Value
35 lb. corn silage	.31	5.79
9 lb. clover hay	.48	3.12
6 lb. corn	.40	5.33
2 lb. cottonseed meal	.70	1.68
	1.89	15.92

This is still a little high in energy and low in protein. It can be made nearer the standard by replacing one pound of the corn with bran: —

	Digestible Protein	Energy Value
35 lb. corn silage	.31	5.79
9 lb. clover hay	.48	3.12
5 lb. corn	.33	4.44
2 lb. cottonseed meal	.70	1.68
1 lb. bran	.10	.48
	1.92	15.51

This ration approaches the standard close enough for practical purposes, although still a little low in protein.

The ration must be modified to meet the requirements of individual animals to some extent, especially regarding the richness of milk produced. The standard as stated is based upon the requirements for average milk. A cow of the Channel Island breeds will presumably require a somewhat higher allowance for the milk, while one of the Holstein breed may use rather less. An exact agreement with the requirements is not essential, even when the milk produced is of average composition, since the composition of the feeds varies to some extent, and the individual requirements of the animals are also subject to some variations.

The Cost of the Ration. — In the foregoing no attention has been given to the relative cost of the feeds used in making up the ration. This question is one of great importance, and must always be taken into account. In general the tendency is for the mill products, such as bran, oil meal, cottonseed meal, and the great variety of others on the market to be sold according to their comparative protein content. In preparing the ration the cost should be calculated at the same time, and the various combinations tried that offer to reduce the cost.

Palatability of the Ration. — It is of considerable importance to take into account the palatability of the ration as well as its composition. An animal will give better results if it relishes its food. Sometimes, on account of a lack of palatability, the cow may not consume as much as she really could use. Hay and other coarse feeds show the most variation in palatability depending upon how they are cured and the stage of ripeness at cutting. It is advisable to have the grain ration composed of a mixture of feed stuffs, as this adds to the relish with which it is eaten. The roughness should consist of at least two varieties. Succulent feed, such as silage and roots, is especially palatable, and aids digestion by keeping the cow in good physical condition. When a good ration is once selected, there is no advantage in making a change for the sake of variety. It has been claimed by some practical feeders that a change in ration is beneficial, but most of the most successful herdsmen of dairy cattle select the ration carefully, then make as few changes as possible.

Order of Feeding. — Regularity in time and manner of feeding is of more importance than any definite order of feeding. As a rule about half of both concentrates and roughness should be fed at night and the remainder in the morning. The grain is usually fed first and the hay feeding reserved until after the milking is completed, to avoid filling the air with dust, which serves to contaminate the milk. Silage should be fed immediately after milking, to prevent the odor from gaining access to the milk. The cow is a creature of habit, and the same routine should be followed. She may be taught to demand her grain ration when milking, but will milk just as well if always fed either before or after milking, and will not look for it at the time of milking.

FEEDING HIGH-PRODUCING COWS FOR THE MAXIMUM PRODUCTION

The maximum production is secured from high-producing cows only by a combination of the expert herdsman and the best possible ration and conditions. Such cows cannot be fed entirely by any rule, nor their ration calculated by a formula. The individual animal and her characteristics must be taken into account as well. One of the essential things is having the animal in the proper condition of flesh at calving. She should be dry for two months or more for the best results, and well fed during this period. Some form of succulence is absolutely necessary as a part of the ration. Roots, such as common beets, sugar beets, or mangels are even better than silage for this purpose, and may be fed up to 50 pounds or more per day.

The cow must be brought up to the full ration carefully after calving, using at least three weeks for this purpose. The grain ration should consist of a mixture of several concentrates all of which are palatable. As long as the animal remains in normal condition, no change in the grain ration is necessary. Special attention must be given to the physical condition of the cow. In this connection the careful herdsman always observes closely the character of the dung excreted, and learns to judge in this way when the digestion is normal. At the first indication of the lack of a keen appetite the ration is cut down until the animal is in a condition to again utilize the full amount. If the digestion shows indication of even slight disorder, a purgative, such as Epsom salts 1 to 1½ pounds at a dose should be administered at once. The grain should always be eaten with a relish, and the animal should show a disposition to want a little more than she receives.

A ration for a very heavy-milking cow must be rich in protein. Much more grain will also be fed in proportion to the roughness than with an ordinary producer. In fact, for the maximum production of a great producer, the nutrients will need to be largely supplied by the concentrates.

The following ration was fed to Bessie Bates, a Jersey cow owned by the University of Missouri. Her production at the time this ration was given was 40 pounds of milk and 2 pounds of fat per day. Her weight was about 900 pounds, and the total product for the year 13,895 pounds of milk and 680 pounds of fat. The same mixture of concentrates was fed during the greater part of her lactation period.

Ration	Lb.
Corn silage	15
Alfalfa hay	15
Corn meal	3.5
Bran	3.5
Oats	3.5
Oil meal	1.5
Total roughness per day	30
Total grain per day	12

Missouri Chief Josephine, a Holstein cow weighing 1350 pounds, received the following when producing an average of 100 pounds of milk per day:—

RATION	LB.
Corn silage	15
Alfalfa hay	20
Dried beet pulp	4
Corn meal	6.1
Bran	6.1
Oats	6.1
Gluten feed	1.9
Linseed meal	1.9
Cottonseed meal	1.9
Total pounds roughness	35
Total pounds concentrates	28

The grain ration was prepared by mixing 100 pounds each of the corn, bran, and oats, and 30 pounds each of the last three named. The cow was fed four times during the twenty-four hours. One pound of dried beet pulp was added to six pounds of the grain mixture, and the entire mass moistened with water some time before feeding.

DISCUSSION OF COMMON FEED STUFFS

No particular food or combination of feeds is alone essential to the most economical production of milk. The first consideration is to grow the most suitable crops on the farm in order that the amount purchased may be as small as possible without reducing the efficiency of the ration. In the brief discussion which follows, only the most common feed stuffs are considered.

Timothy Hay. — The value of this hay is often greatly overestimated as a feed for dairy cows. It is unpalatable except when cut early, and will not be consumed in sufficient quantities. The most serious objection is the low protein content, making it necessary to feed large quantities of concentrates rich in this expensive nutrient when timothy hay is fed.

Corn Stover. — This forage may be utilized to a small extent. It has the same characteristics and objections as timothy hay, and cannot be depended upon for more than a part of the roughness.

Hay from Legumes. — Hay of this class is especially valuable for the dairy cow. It includes the common clovers, alfalfa, the cowpea, the soy bean, the field pea, and other less commonly used legumes, such as vetch and crimson clover. Forage from this class of plants when properly cured is highly palatable, and contains a relatively large amount of protein. It is for this reason especially that a legume hay should by all means be grown by the farmer in the corn belt. The ash content is also large, which is of importance, especially when fed with corn products that are low in this class of substances.

Silage. — The importance of supplying a succulent food to the cow at all times has been discussed elsewhere. In feeding corn silage it should be kept in mind that it is not of itself a complete ration for the cow in milk, since it is relatively high in carbohydrates and low in protein. It is not advisable, either, to feed it as the only roughness. Some hay should be given as well, and for this purpose the legumes are the best adapted, on account of their high protein and ash content. It is not advisable to feed over about 35 pounds to a small cow and 40 to 45 to a large animal.

Corn. — Over the greater part of America corn is the most common and cheapest grain. In the corn belt this valuable grain is often fed to excess. On the other hand, some dairymen avoid the feeding it altogether, on account of the erroneous idea that it is not suited to a cow producing milk. Corn may be fed in reasonable quantities to any class of animals on the farm. It is especially palatable for the cow in milk. However, it must not be the exclusive grain ration for good results. The protein content of corn is low, likewise the ash. If combined with corn stover or corn silage for roughness, the protein content is entirely too low for a dairy ration. Corn silage and ground corn combined with clover or alfalfa hay and bran, however, makes a good ration for general feeding. It should not be used in excess for the growing or pregnant animal.

Wheat Bran. — Next to corn wheat bran is the most important cow feed of this country. Its great value as a food for growing animals and cows in milk comes from the high ash and protein content. Its light, loose character also makes it a valuable addition to a heavy ration in the way of lightening up the mass so it is easier acted upon by the digestive juices. This is of special importance in connection with such feeds as cottonseed meal that have a tendency to form a pasty mass in the stomach which is difficult to digest.

Wheat middlings or shorts are valuable feeds for the cow, but more like corn meal in composition and properties than like bran. As a rule it is wiser to make use of the bran rather than shorts for the cow in milk.

Oats and Oat Products. — Oats are a splendid feed for cows and growing animals when the cost is not prohibitive. Woll found oats to be about 10 per cent more valuable pound per pound than bran when fed to cows. In general, it may be said that oats are themselves an excellent ration, but do not contain sufficient protein to be as effective in supplying a deficiency in that respect as are others with a high protein content. The valuable by-products of oats are mainly from oatmeal mills, and consist of oats shorts and finely divided parts of the grain sifted out. In addition, a much larger quantity of hulls must be disposed of by these mills. Hulls are mostly crude fiber and are hardly equal to the same weight of timothy hay in feeding value. The by-products of the oatmeal mills are therefore valuable, to the extent that they contain the parts of the grains. Oat hulls are used largely to form a portion of various mixed feeds that are put upon the market.

Cottonseed Meal. — This by-product is the residue left after the oil is extracted from the cottonseed. It contains the highest amount of protein of any feed used for cows ordinarily found upon the market. For this reason it is especially valuable as a means of balancing up rations deficient in protein, where corn and corn products form a large proportion of the ration. It should not be fed to excess at any time. As a rule from two to four pounds per day are to be considered the maximum to be used. However, in the South, where it is abundant, it is fed in much larger quantities with good results.

Linseed Meal. — This valuable feed is the residue after the linseed oil is extracted from flaxseed. It ranks next to cottonseed meal in protein, but on the market usually sells a little higher. It seems to exert a specially favorable effect upon animals of all kinds to which it is fed. Like cottonseed meal, it is especially valuable as a means of supplying the protein liable to be lacking in the farm-grown ration.

Gluten Feed. — This is a by-product from starch and glucose factories. It consists of the corn grain after the starch is extracted. In protein content it ranks about midway between bran and oil meal, and is a palatable and valuable feed.

Beet Pulp and Molasses. — Since the beet sugar industry has become of some importance in the United States, the by-products are found on the market for cattle feed. The beet pulp from the factory contains only about 10 per cent of dry matter. On account of its bulk it cannot be transported any distance, and is fed only in the immediate neighborhood of the factory. More recently the dried pulp has been placed on the market and is meeting with favor as a feed. In composition it is high in carbohydrates in proportion to the protein, ranking in this respect below corn. In feeding it should be combined with other feeds richer in protein. It swells when moistened, and cannot be pressed into a compact mass. For this reason it is easy of digestion and valuable to lighten up a grain ration that otherwise would form a mass in the stomach not easily penetrated by the digestive juices.

Another refuse or by-product of the cane and beet sugar factories is low-grade molasses. This substance is sold now in combination with a variety of other feeds, such as beet pulp, alfalfa hay, and sometimes with worthless material, such as peanut hulls, weed seeds, or cocoa wastes. Molasses serves a useful purpose as a means of making unpalatable feeds more readily consumed. Unfortunately it is too often used to cover up inferior quality or to disguise material that is of little or no feeding value. The general advice regarding feeds of this class is to purchase them only on the advice of the experiment station in the state where the product is sold.

Brewer's Grains. — Fresh brewer's grains are fed in large quantities, where they may be hauled directly from the brewery. Considerable objection has been raised by city health authorities in many places to the use of this feed. If fed in moderate amounts under proper sanitary conditions, they are not objectionable. However, the use of them is so often abused that some officials have found it easier to prohibit their use than to regulate it. The objection comes from feeding them exclusively, from allowing decomposition to begin before feeding, and from the very objectionable sanitary condition that exists if special care is not taken to keep the feed boxes, feeding troughs, and, in fact, the entire stable, clean.

This feed should not be used to exceed twenty pounds per day, and should be supplemented with hay and some other grain, such as corn. The greater part of the brewer's grains now produced are dried, and in this form may be transported long distances. They are rich in protein, and four or five pounds may be used in the ration with advantage. At present the larger part of this by-product finds a market in Europe.

Mixed Feeds. — No small proportion of the grain supplied the dairy cows of the United States is in the form of mixed feeds. As a class any mixed feed is to be looked upon with suspicion. Where the unmixed grains and by-products are to be had on the market, it is always safer to purchase them and make such mixtures as may be advisable for the purpose in view. The main purpose in view by the manufacturers or dealers in preparing feed mixtures is to sell material of inferior quality or some by-product of little or no value. One of the most common ingredients of mixed feeds is oat hulls, from oatmeal factories. In many cases the hulls are ground fine to escape detection, while the claim may be made that ground oats is a part of the mixture. A careful examination will usually indicate if oat hulls have been added. Wheat bran is occasionally mixed with ground corn cobs or corn bran.

A cottonseed feed is also found upon the market which is a mixture of cottonseed hulls and cottonseed meal. The only object in making such a mixture is to sell as much cottonseed hulls as possible at a good price. Alfalfa hay of doubtful quality is mixed with sugar refuse, and by liberal advertising sold at a price above its real value.

Nearly all states where large quantities of feed are purchased by the farmers now have some law in force regarding the sale of feeding stuffs. These laws, however, do not take the place of intelligence on the part of feed users. Such a law generally requires the proper branding of each sack and labeling to indicate the chemical composition. It should be remembered that the label gives the total amount of protein and other constituents, and not the amount of each that is digestible, which will be decidedly lower. Every feed buyer should patronize only reliable dealers, and buy feeds that are labeled and guaranteed. There are no mixtures better than the buyer can make himself, and there is no special feed or mixture having any remarkable properties not possessed by familiar feeds. The buyer of mill feeds should make a point of keeping in touch with the experiment station of his state, and if the feed control is vested in some other body or official, with them as well, and make use of the information they will be able to furnish regarding the feeds on the market.

CHAPTER XXIII

STABLES FOR COWS

THE dairy cow, unlike the fattening steer that is protected by layers of fat, needs to be comfortably housed to do good work. The loss from exposure to cold, and especially cold rains, results in much larger losses than the actual amount of feed required to maintain the animal heat under the unfavorable conditions. The importance of housing is generally understood and practiced in the colder climates. As a rule more losses from exposure occur in those regions where, from the usual mildness of the climate, sufficient provision is not made for the severe weather that occurs only at intervals.

Where the importance of proper stabling is recognized, the conditions existing are, on the average, far from what they should be. No part of the present system of handling cows is more in need of improvement at present. It must be recognized that the stable is a place in which human food is produced, and further that the health and even the lives of the children of the country depend to a large extent upon the conditions existing in the stables where the milk, which serves as their main food, is produced. There is a strong and growing demand on the part of milk consumers and officials having to do with the health conditions of the cities for better sanitation in the barns and dairies. Laws and regulations regarding this matter are becoming more stringent, and it is safe to say that an immense improvement will be brought about within the next few years.

Better Barns mean Cheaper Production. — There is another phase of the subject that must be emphasized as well. There is no doubt it pays as a financial proposition to have well-arranged, sanitary barns making the cow more productive by being more comfortable; labor more efficient on account of being better satisfied; and the expense of labor less on account of convenience in arrangement. A sanitary barn is not necessarily an expensive one. Many an inexpensive structure is or may be more sanitary than some ill-arranged, badly kept, but expensive barn.

In considering the matter of barns, it is well to study carefully how closely the conditions of early summer may be maintained throughout the year. The dairy cow does her best in the early part of the summer when on a good pasture. The maximum production reached at this season is possibly largely on account of the excellence of the food, but at the same time the animal enjoys a moderate temperature and clean, comfortable surroundings. There is an abundance of fresh air and sunlight, and the cow has perfect freedom of movement. Keep these conditions as near as possible in the barn, and good results will follow. A cow kept in a dark basement barn, surrounded by foul air, with her head fast in a rigid stanchion and her body more or less filthy, is as far from summer conditions as is her milk production below that of early summer.

Types of Barns. — There are a variety of barns, but they may be divided into a few rather distinct types, while many are intermediate.

296 DAIRY CATTLE AND MILK PRODUCTION

1. The basement barn.
2. The two-story or loft barn.
3. One-story or shed barn.
4. The round barn.
5. The covered barnyard or double stabling system.

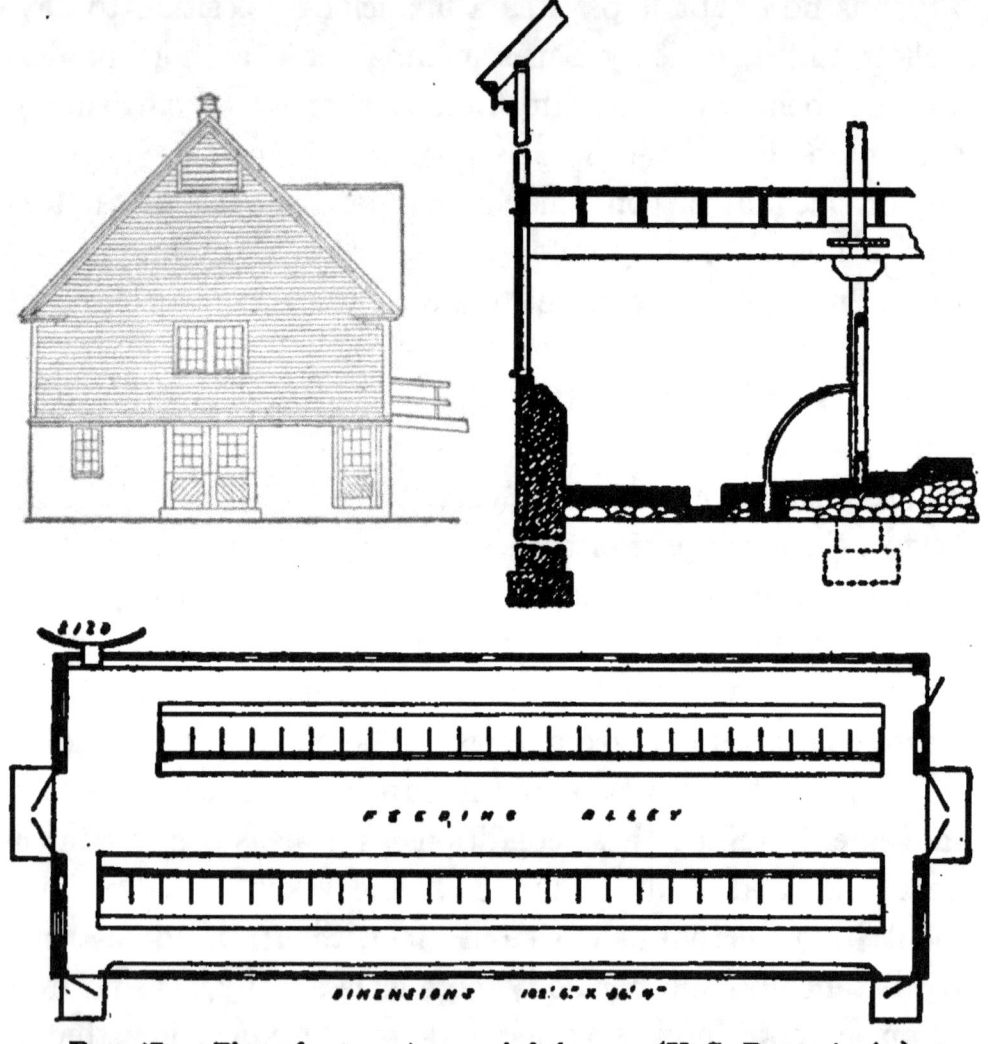

Fig. 47. — Plan of a two-story or loft barn. (U. S. Dept. Agric.)

The Basement Barn. — This is a favorite type in the northern part of the United States, especially in the Eastern States. It cannot, by any means, be recommended as an ideal dairy barn. It is built by excavating into the side of a hill, sufficiently to bring the top of the first story on one side, and usually on two sides, at the level of the ground outside, the south and east sides commonly being full height above ground. This type of barn is warm, but usually very unsanitary, on account of having practically no light and no ventilation. It can be ventilated, but rarely is this done. Space for storage is usually provided above the animals.

Two-Story or Loft Barn. — This style of barn allows for the stabling of the stock on the first floor, with a second story for storage above. The walls of the first story may be of stone or wood, but are all above ground. This style is well adapted to the general farm, where considerable room is required to store the loose hay and other feeds grown on the farm for winter use. Storage room is secured more cheaply in this manner than by building a one-story cow barn and a separate storage barn. This style may be entirely sanitary in its construction if properly arranged. It should have plenty of light and a good ventilating system. The ceiling above the cows should be tight, to prevent dust from falling, which would contaminate the milk, and to avoid the odor from the stable and the breath of the animals from injuring the palatability of the feed above.

The One-Story or Shed Barn. — This is one of the best types of barn from the standpoint of sanitation and convenience. It usually is built wide enough for two rows of cows. This plan is especially adapted for use where it is not necessary to have any large amount of storage room for unbaled hay or bedding. It is often used, however, where ample storage room must be provided, and in such cases a portion of the barn is usually built two stories high, and serves for storage and general purposes, while one or more single-storied wings are provided to house the cattle. Single-storied barns usually have a tight ceiling, although some are open to the roof and have a monitor top to admit light from above. The objection to having the space open to the roof is that in cold climates this space is so great that the barn is too cold in winter.

The advantages of the one-story barn are that it may be well lighted and ventilated and the construction made sanitary in every way. Additional room can be easily had by extending the wings at any time.

The Round Barn. — The economy of construction in the round barn was first called to attention by King.[1] According to this author, a round barn requires about 25 per cent less wall to inclose it than does an ordinary rectangular type.

According to Frazer,[2] a round barn with room for 40 cows requires 22 per cent less walls and from 34 to 58 per cent less material than a rectangular building with accommodations for the same number of animals. The silo is built in the center, and the cows are usually arranged in a single row around the barn, headed toward the center, except where the barn is too large, when a double row is used.

The Covered Yard or Double System of Housing. — This plan was first called to the attention of the public by Professor Roberts[3] of Cornell University, who used it for many years with the station herd of dairy cows. The plan consists in having a large shed or covered yard, into which the cattle

---

[1] *Physics of Agriculture*, p. 366.
[2] Bulletin No. 143, Illinois Experiment Station.
[3] Bulletin No. 13, Cornell University Experiment Station.

are turned loose except at milking time. The roughness is usually fed in racks to be consumed at will, while the grain is fed when the cows are taken to the milking stable at milking time. The milking stable need not be elaborate, but sanitary. The cows are tied while they eat their grain and are milked. This system gives the cows a maximum amount of freedom, but requires an abundance of straw for bedding, and is more expensive if a new and sanitary plant is built. At times an old barn may be utilized in this manner economically.

Location of the Barn. — The barn should be located where there is good drainage, making it possible to keep the yards in good condition. Convenience in location as well as in planning should receive careful consideration, since the expense for labor depends to no small degree upon proper location and internal arrangements.

A rectangular barn should stand preferably north and south, making it possible to get sunlight on both sides at some time during the day.

Lighting. — One of the most serious defects in most barns, especially the older ones, is a lack of sufficient light. Windows cost but little, if any, more than other wall space, and cannot be objected to on this account. Plenty of light is one of the most essential things about a good barn. It is necessary to keep the animals in a healthful condition, and is of the greatest importance in a sanitary way. Not only does sunlight destroy germs, but by having plenty of light uncleanly conditions are easily seen and corrected. A dark barn is almost always a dirty barn. There should be at least 4 square feet of glass to each animal. The bottom of the windows should be 4 or $4\frac{1}{2}$ feet from the floor. If lower, the animals are liable to break the glass. The window should extend nearly to the ceiling in order to allow the sunlight to reach as much of the floor as possible. The windows should be set flush with the inside wall to avoid making a ledge for the accumulation of dust. In a one-story barn the windows should be hinged. In a two-story barn, which gives ample room above, the sash had better slide upward. This arrangement is much more convenient than having one sash slide past the other in the ordinary manner.

FLOOR CONSTRUCTION

One of the most important points to be considered in planning a barn is the material to be used for floors. Material for a floor should have the following characteristics: —

1. Impervious to moisture.
2. Sanitary and easily cleaned.
3. Comfortable for the cows.
4. First cost not too great.
5. Durable.

The floors in common use are as follows: —

Dirt with wood or cement gutters.
Wood.
Brick.
Cement or granitoid.

Dirt Floors. — A floor of dirt in a dairy barn is only excusable under primitive conditions. This material provides comfort for the animal, and is cheap in construction. The objection is, of course, that it cannot be kept clean. The most objectionable condition is where the cows stand on a level dirt floor without any gutter. A fairly good floor for ordinary purposes may be made by building a gutter of

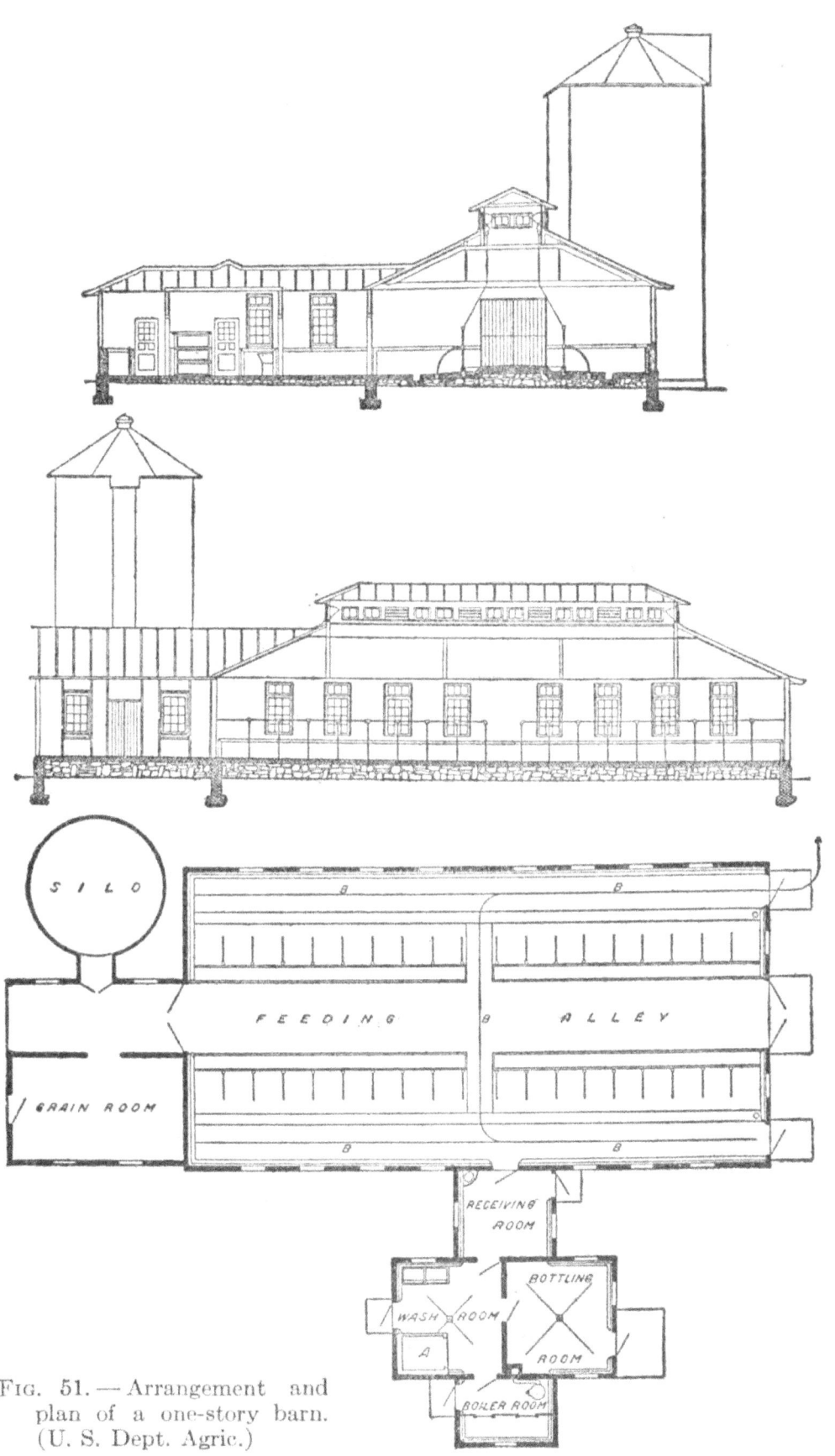

Fig. 51.—Arrangement and plan of a one-story barn. (U. S. Dept. Agric.)

cement or wood, and extending this forward far enough to catch the urine. The space under the cows is then filled with clay packed solidly. When sufficient bedding is used, such a floor will answer in a cheap barn, but under no circumstances could such a barn be called sanitary.

Wood for Floors. — A tight wooden floor is comfortable for the animals, and may be kept in good condition regarding cleanliness, although it can hardly be considered first class from a sanitary standpoint. The first cost is also excessive, considered in connection with the short time it remains in service. Wooden floors last the longest either laid in contact with earth so moisture is retained constantly, or laid with sufficient air space below to admit of free circulation of air. Under the most favorable conditions a wooden floor may last as long as from six to ten years. The most rapid decay occurs when the floor is laid far enough above the soil and with no circulation of air underneath, so only a small amount of moisture is present. Under such conditions the floor may not last over three to five years.

Wooden floors are made water-tight by using coal tar between the planks. The most serious objection to the wooden floor is its short period of service. Another objection of considerable weight in many cases is the sanitary question. On account of the difficulty of cleaning a wooden floor, it is not used where the greatest attention is paid to sanitation.

Brick for Floors. — If good vitrified brick can be bought cheaper than cement, this material may be used with advantage. A brick floor has the same advantages and objections as discussed in connection with cement. The bricks must be put on on a good foundation, and are set in cement.

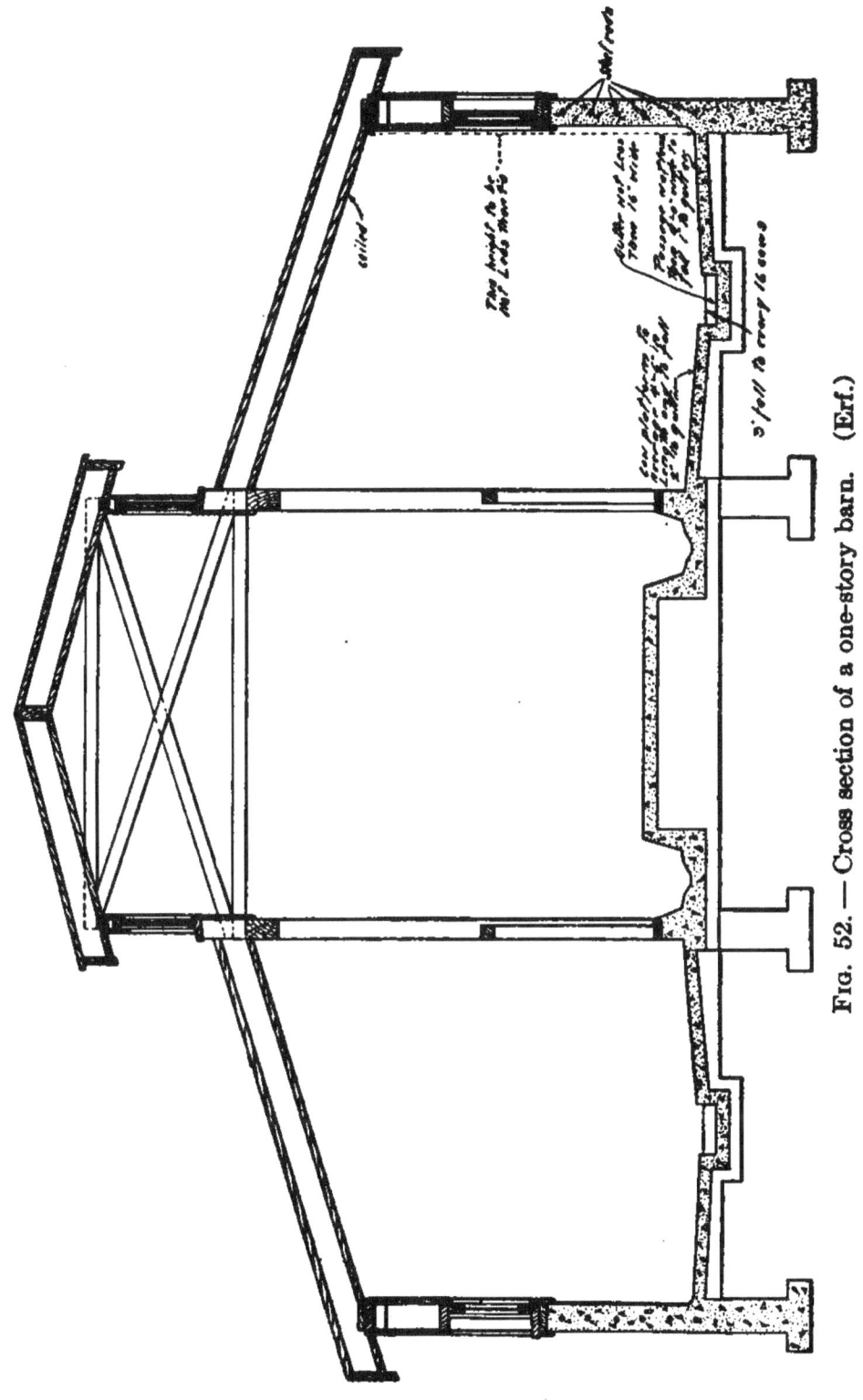

Fig. 52. — Cross section of a one-story barn. (Erf.)

Cement Floors. — Cement or concrete floors have more of the requirements for a good floor than any other material in general use. This material is impervious to moisture, very durable, and the most sanitary and easily cleaned of all. The first cost is but little more than wood in most localities, and its lasting properties make it much cheaper, considering a term of years. The one serious objection to cement is in regard to the comfort of the animal. A cement floor is cold, that is to say, it is a good conductor of heat, and for this reason seems cold. It is believed by many of the most experienced dairymen that udder troubles are brought on from cows lying on cement floors. This difficulty may be lessened by having the floor 6 to 9[1] inches higher than the surface of the ground on the outside to prevent water flowing under the cement. The floor should be thoroughly insulated with a layer of cinders 6 to 9 inches deep under the cement. Such arrangement, together with a liberal use of bedding, will obviate most of the danger from the coldness of the cement. Another arrangement which has much to recommend it is placing a wooden platform over the cement where the cows stand. This platform may be made of planks imbedded in coal tar, or in the form of a loose frame that may be removed for cleaning.

Another serious objection to cement is the slipping of the animals as they come through the passageways. A cement floor should never be trowled to a smooth finish, but left with the surface rough as finished with a board. In some barns sand is sprinkled on the floor daily to prevent the animals from slipping. Another trouble that often occurs is injury to the knees of the cows. In reaching in the manger for food

[1] Erf., 15th Annual Report Ohio State Dairy Association, 1909.

their forefeet slip and they drop on their knees, resulting after a time in enlarged joints. This trouble is avoided in a large measure by making a depression of an inch where the forefeet stand.

Probably the best plan is to construct the entire floor of the barn, including stalls, passageways and mangers, of cement, then to cover the platform where the cows stand, as suggested, with wood.

Arrangement of Cattle in Barn. — The best and most common plan is to stand the cattle in a double row. This makes it possible to light the entire stable readily and to feed and remove the manure conveniently. There is much discussion as to the comparative advantages of placing the cows with their heads together or with heads outward. There are some advantages in favor of each. When facing the center, both rows may be fed by making one trip down the passageway with the feed truck, distributing the feed on both sides. The removal of the manure is most convenient when the animals are headed out, since by driving through with the manure spreader and loading directly from the trenches, considerable labor is saved, or if a track carrier is used, it may be loaded readily from both gutters. The cows present a better appearance to visitors when headed out. The walls of the barn do not become splashed with manure with this arrangement. In regard to ventilation, there is some advantage in having the heads outward, since the fresh air usually enters along the outside wall. The greatest difficulty that is usually experienced in heading the animals outward is constructing the barn without the use of center posts that are in the way and obstruct the passageway behind the animals. When the cows are headed toward the center, the posts for support are

x

made part of the stanchion manger supports, and are not in the way. It is possible, however, with more expense, to support the longer span required with trusses when the cows are headed outward. Figure 53 shows cross sections of barns conveniently arranged. The width of the barn is 36 feet. It will be observed that the cement comes up on the wall 3 feet, and is made with a rounded corner to prevent accumulation of dirt. The passageways should be of ample width to facilitate feeding and cleaning out.

Gutters. — The gutter is highly important in connection with keeping the cow clean. It should be of ample depth, as otherwise cows are apt to stand with the hind feet in the gutter. The depth should not be less than 8 inches, while 10 or 12 inches is better. The proper width is not less than 16 inches. The danger from deep gutters

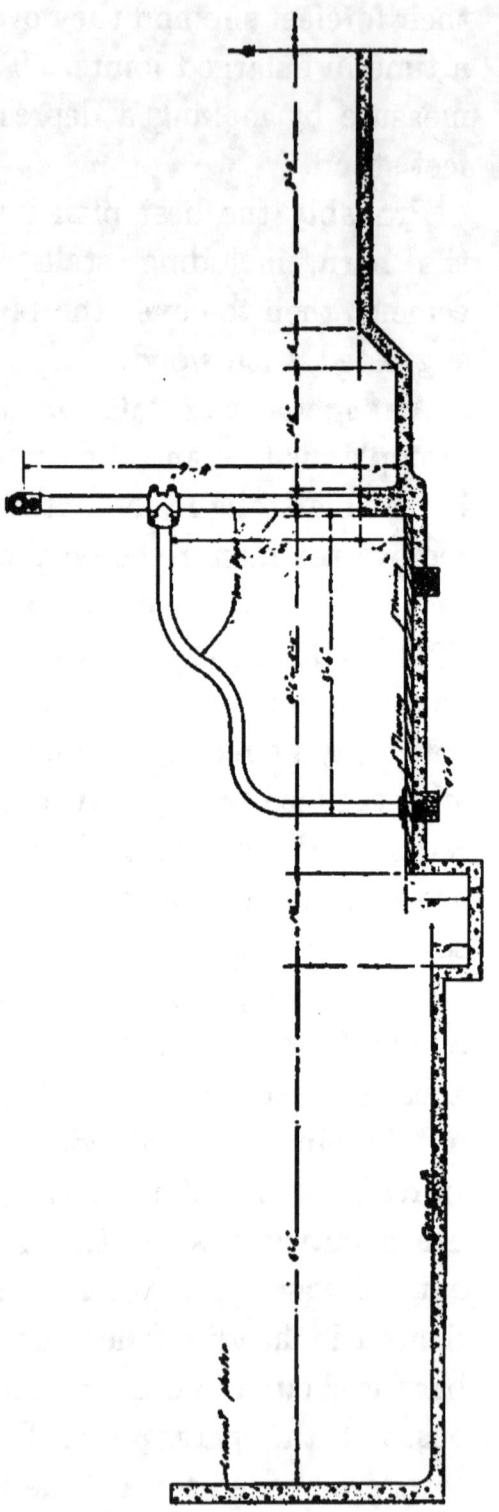

STABLES FOR COWS

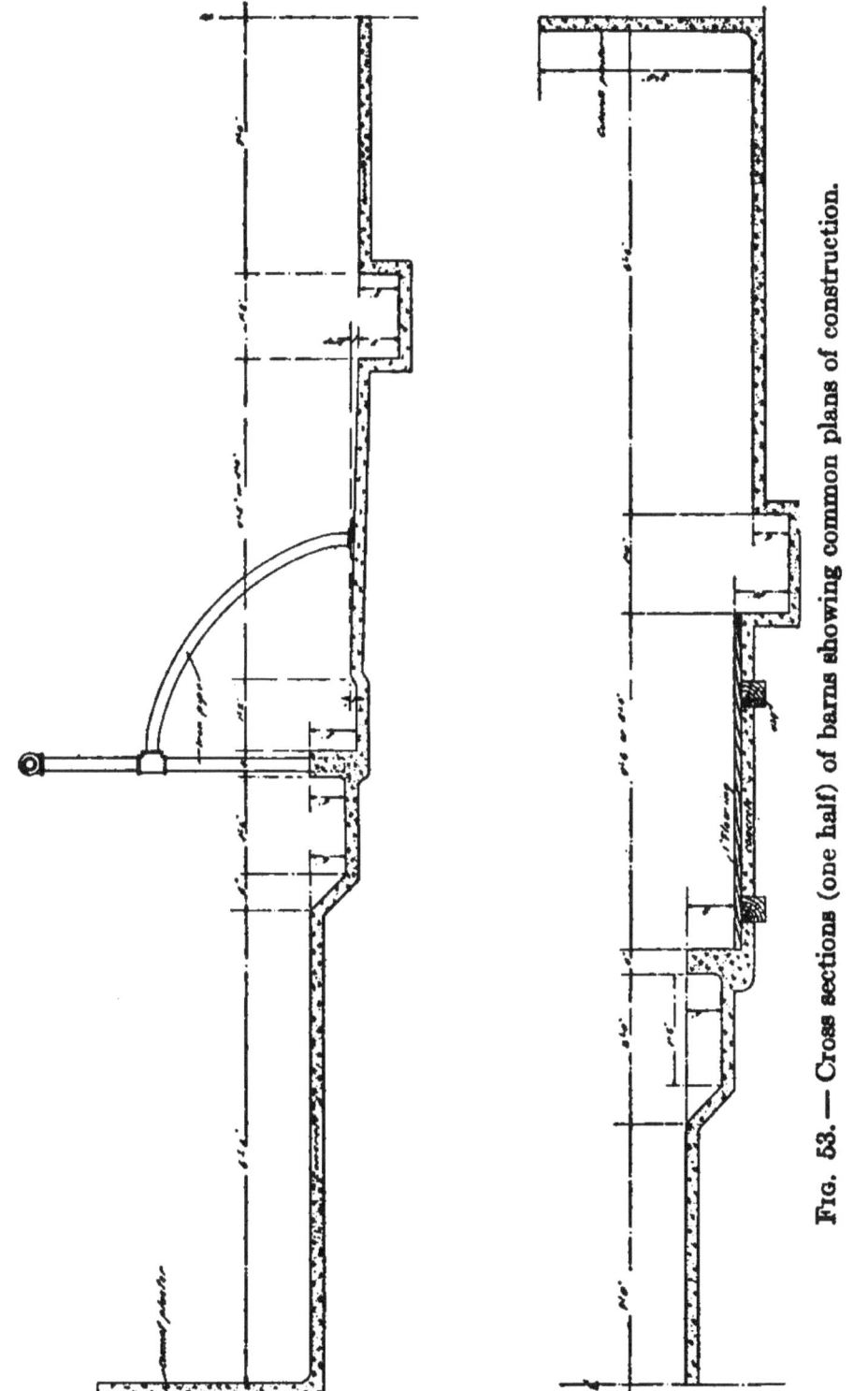

Fig. 53.—Cross sections (one half) of barns showing common plans of construction.

is that the cows will slip and injure themselves. This may be largely eliminated by lowering the passageway behind, making this side of the gutter only 6 inches deep.

Mangers. — There are numerous types of mangers. From a sanitary standpoint cement is the best material for construction. All corners should be rounded to facilitate cleaning. The most common type is the continuous manger, which is built in the form of a long trough before the cows. Some objections are raised to this construction, on account of the chance it affords for one cow to rob another of her feed. In case of contagious diseases like tuberculosis, there is also much more danger of communication from one animal to another in such a manger. Partitions of sheet iron are sometimes used. The special advantage of the continuous manger is the ease of cleaning by sweeping out refuse feed. It may also be used as a means of watering the cows in the barn with advantage.

Fig. 54 shows some of the best types of cement mangers. The width should be not less than 2 feet, and preferably 2 feet 6 inches. The bottom of the manger should be 1 or 2 inches higher than the platform where the cows stand. The partition next to the platform on which the cows stand should be 6 or 8 inches above the level of the bottom of the manger.

The comfort of the cows and the success in keeping them clean depends largely upon the construction of the platform. It is highly important that it be of proper length and of suitable material.

The construction to be recommended most highly is of cement covered with a wooden platform. A cement platform should have a depression one inch deep and 14 inches wide next to the tie. The surface of this depression should

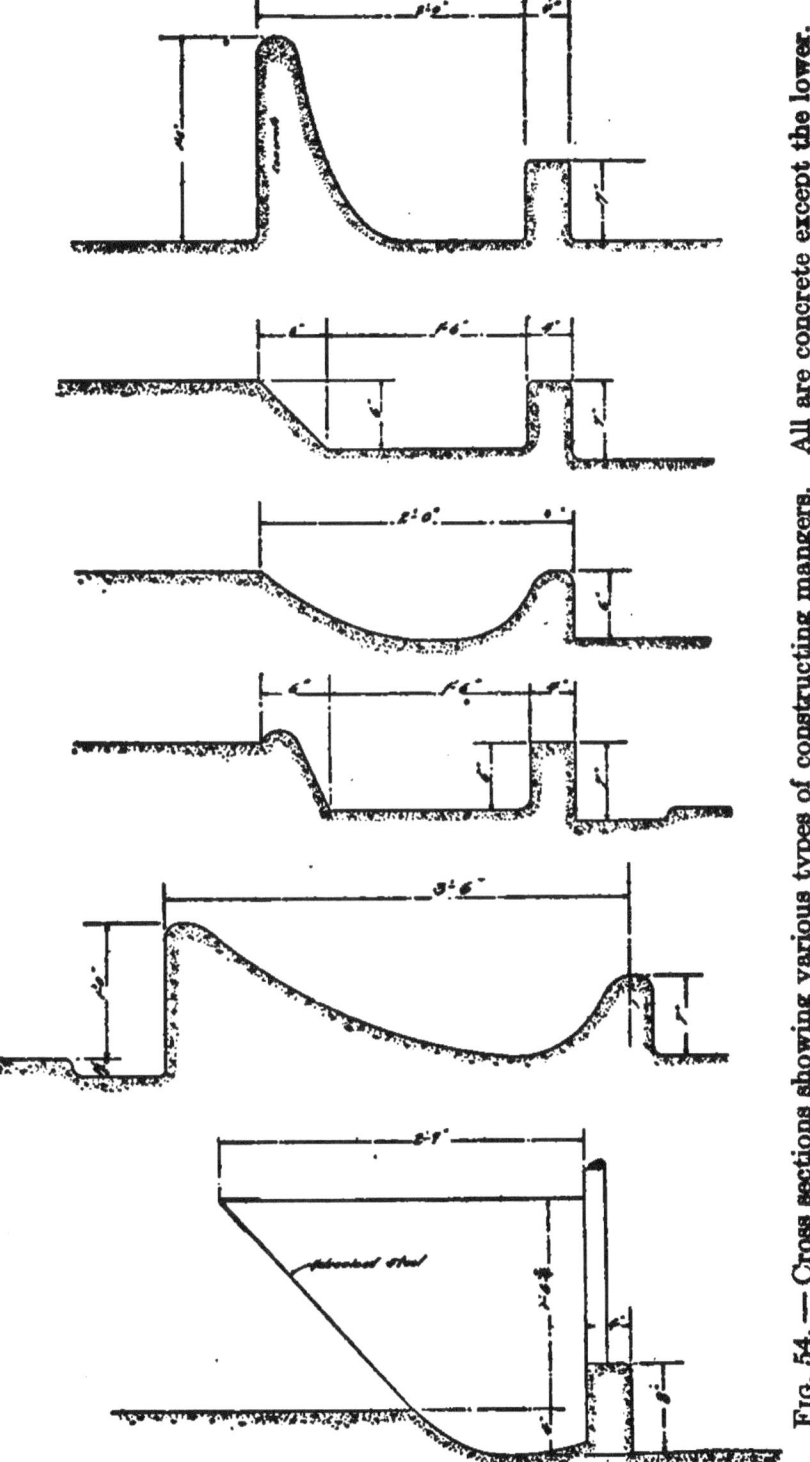

Fig. 54.—Cross sections showing various types of constructing mangers. All are concrete except the lower.

be slightly fluted. The object in this is to prevent the cows from slipping and falling on their knees when reaching for feed. It also allows the cow to stand on the level, since the platform slopes one inch from the rear of this depression to the front edge of the gutter. The success of keeping the cow clean depends largely upon having a platform of proper length. The length to be used varies from 4 feet 6 inches to 5 feet, depending upon the size of the cows. The former is the proper length for Jerseys, and the latter for cows of the size of Holsteins. Some provision should be made to accommodate cows of different size, since they vary in any herd with age. One plan often followed is to make the platform 4 inches longer at one end than at the other, with a gradual slant between. The cows are then arranged according to size. Another arrangement having much to recommend it is an adjustable stanchion so made that it may be set back 3 or 4 inches from the support, or set ahead the same distance for long cows.

Ties. — There are a great variety of ties in use. The most objectionable way to tie a cow is to fasten her to a manger where she must back up to lie down. This is bound to result in filthy animals, since they are compelled to lie in their droppings. The cow should be so fastened that she lies down exactly where she stands, or a little forward if possible, and the platform should be the proper length so the manure drops in the gutter.

The most common ties in use are various forms of stanchions. There is no kind of tie that keeps the cows cleaner than the rigid stanchion, provided the platform and gutter are properly made. The rigid stanchion, however, is not well suited for a tie, as the cow has no freedom of movement and cannot lie in a natural position. Many other forms of stanchions are in use that are quite satisfactory. One is hung on pegs at top and bottom, allowing a movement sideways. Another is hung on chains at top and bottom, which gives more freedom. For general use these improved forms of the stanchion find the most favor. Stanchions may be made of iron pipe or wood, but the former is the more sanitary and more durable, and equally comfortable for the animal.

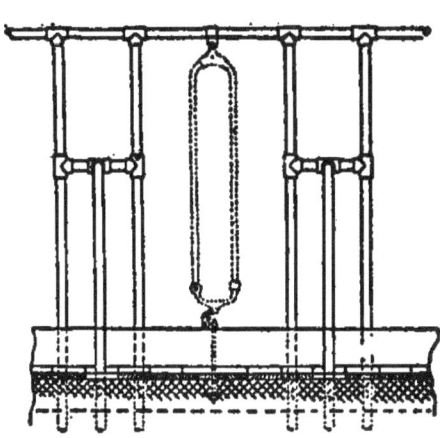

Fig. 55. — Stall constructed of iron piping.

The double-post slip chain tie is equally comfortable for the animal, but not quite so convenient for use.

Stanchion or chain ties may be attached to either iron or wooden framework as supports. The iron is most sanitary, and has the additional advantage of being more sightly. An iron pipe partition made of gas pipe, with the bottom set in the cement floor, is to be recommended.

VENTILATION

An abundance of pure air for animals of all kinds is scarcely less important than are proper methods of feeding. It is only within recent years that the full significance of an abundant supply of oxygen to animal life has been comprehended. While there may be some excuse at times for insufficient or improper feeding of animals, there is none for failure to supply plenty of air.

Good ventilation for dairy barns is not only necessary

from the standpoint of the health of the animals, but it is necessary for the most economical production of milk. Dairymen are largely indebted to Professor King for information on this subject, as well as for the best practical system of applying this knowledge.

The cow gives off carbon dioxide, moisture, ammonia, marsh gas, and some other organic matter from the lungs. A candle will be extinguished if placed in a jar containing the air exhaled from the cow's lungs. A cow weighing 1000 pounds inhales 224 pounds of air in 24 hours, or about double the amount by weight of her food and drink. This is at the rate of 3542 cubic feet per hour. To supply this amount of air for 20 cows will require a ventilating flue 2 by 2 feet, in which the air moves at a velocity of 295 feet per minute. In providing pure air for stables the cubic space per animal has little significance. The important question is the amount of fresh air provided.

Forces producing Ventilation. — There are three main forces that cause movement of air in a stable.

1. The wind pressure against the side of the building, which tends to force air into the building and out on the opposite side or upward through the ventilators.

2. The wind in blowing across the top of a ventilating flue produces an outward suction.

3. The difference in temperature between the air in the barn and that on the outside. This causes an upward movement in a ventilating shaft by a force equal to the difference in weight of the air outside and within.

The King System. — This is the system in general use in modern dairy barns. The main features consist of a large flue opening near the floor and extending above the roof of

STABLES FOR COWS

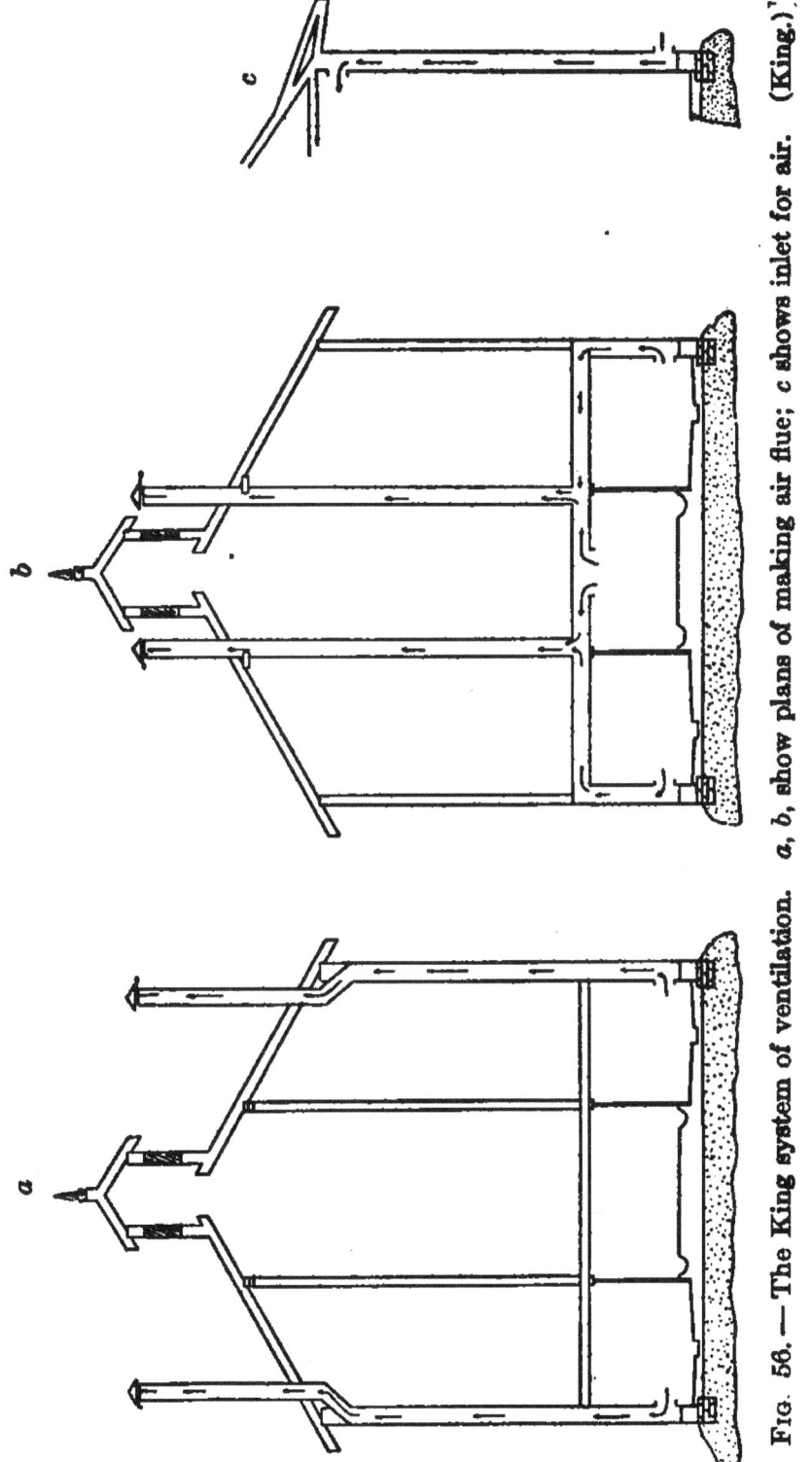

FIG. 56.—The King system of ventilation. *a, b,* show plans of making air flue; *c* shows inlet for air. (King.)

the stable for the escape of the air. A series of smaller openings are arranged on either side for the air to enter.

The object in taking the air from near the floor rather than at the ceiling is to remove the coldest air and the most impure. The warmest and the purest air is found at the ceiling. The ventilating flue should be smooth inside, practically air-tight, and, for good results, with no turns. The flue must have ample cross section. If too small, the friction is sufficient to prevent free movement of the air. In many cases poor results in using this system are to be credited to a ventilating flue of too small size. None should be built less than 2 by 2 feet.

The ventilator flue should have an opening near the ceiling that may be opened when it is desired to increase the draft, and in warmer weather when there is no reason for conserving the heat of the stable. This opening should be regulated with a register.

Entrance for Fresh Air. — Provision must be made for the entrance of outside air. This is taken in at the ceiling, and mixes with the warm air. The intakes should extend downward in the wall, with the opening to the outside three feet or more lower than the opening in the barn. This is to prevent the warm air in the stable from flowing out. These intakes should be on all sides of the barn, to take advantage of all wind pressure. They should not be over 4 or 5 by 16 inches in size, and provided with registers to regulate the air passage.

The King system works only while the stable is closed. At such time as it is not desirable to close the barn, the air should be allowed to enter the ventilating flue through the opening near the ceiling. The King system cannot be expected to work in a barn not tightly constructed so the air finds entrance or exit at other places. Where properly installed, this system gives excellent results. A barn filled with animals will show no barn odor in the morning after being closed over night where this system is used.

Window Ventilation. — Some well-constructed barns depend upon hinged windows for ventilation. These are usually hinged at the bottom, so the top may be tilted inward to the desired extent. By careful constant attention this plan may give fair results. The movement of air in this case is dependent upon wind pressure and further the warmest and not the coldest air is removed.

CHAPTER XXIV

HANDLING MANURE; MATERIAL FOR BEDDING

Composition and Value of Manure. — In all countries where agriculture has been highly developed, the value of barnyard manure is fully appreciated. It is saved with great care, and applied to the soil under the best conditions possible. In purchasing feeds, the probable fertilizing value is taken into account, as well as the feeding value.

Experiments at the New Jersey and Pennsylvania Experiment stations have shown that on the average per 1000 pounds live weight, the cow excretes 46 pounds of dung and 27 pounds of urine, a total of approximately 70 pounds per day, exclusive of bedding. This amounts to over 12 tons per year without bedding, and from 14 to 15 tons with bedding.

According to Professor Snyder of the Minnesota Experiment Station, barnyard manure gives a return of from $2 to $3 per ton when applied to the soil, the latter amount being realized when manure is applied to a soil reduced to a low state of fertility. This value is computed by actual increase in crops produced in five years' time in tests made by the Minnesota Experiment Station. At this rate the manure from a single cow would be worth from $25 to $30 per year. This value is practically the same as the estimate made by the Cornell Experiment Station, based upon the analysis of barnyard manure as given in the following table: —

AMOUNT AND VALUE OF MANURE PRODUCED PER 1000 POUNDS OF LIVE WEIGHT OF DIFFERENT ANIMALS

	AMOUNT PER DAY	VALUE PER DAY	VALUE PER YEAR
	Lb.	Ct.	$
Sheep	34.1	7.2	20.00
Calves	67.8	6.7	24.45
Hogs	83.6	16.7	60.88
Cows	74.1	8.0	29.27
Horses	48.8	7.6	27.74

The distribution of the fertilizing elements, nitrogen, phosphorus, and potash, contained in the ration-fed dairy cows was found by the Pennsylvania Experiment Station to be as follows: —

	NITROGEN	PHOSPHORIC ACID	POTASH
	Per Cent of Total Excretion	Per Cent of Total Excretion	Per Cent of Total Excretion
Dung	31.14	75.55	15.58
Urine	52.33	1.42	74.56
Milk	16.53	23.03	9.86
Total	100.00	100.00	100.00

Roughly speaking, the feces contained one third of the nitrogen and three fourths of the phosphoric acid excreted, while the urine contained one half of the nitrogen and three fourths of the potash.

It will be seen that the milk contained a total of 16.5 per cent of the fertilizing value of the food. The remainder of the plant food under proper conditions finds its way back to the soil. If butter or cream is sold, and the skim milk retained on the farm, practically no fertilizing material is removed from the farm, since butter fat has no manurial value.

It is a fact often lost sight of in practice that the urine of animals contains by far the most valuable fertilizing constituents of the excreta. In the investigations of the Pennsylvania Experiment Station, to which reference has been made, it was found that more than one half of the manurial value of the food and a total of 63 per cent of the manurial value of the excreta was in the urine. If it is allowed to run to waste, as is often done, the larger proportion of this plant food is lost. Practical means of handling and preserving the urine is one of the difficult problems connected with the management of a dairy barn. Another point that comes into any general consideration of the subject is the proper management of the manure and the losses that occur from leaching and fermentation. The New Jersey Experiment Station found that when solid cow manure was exposed to ordinary leaching for 109 days, it lost 37.6 per cent of its nitrogen, 51.9 per cent of its phosphoric acid, and 47.1 per cent of its potash. Mixed dung and urine lost in the same time 51 per cent of its nitrogen, 51.1 per cent of its phosphoric acid, and 61 per cent of its potash. Over one half of the total value was lost in less than four months' exposure to ordinary weather. The loss in fertility from one cow by the leaching of the manure would amount to $12.50 per year. It would add 25 cents per 100 pounds to the cost of milk from cows producing 5000 pounds per year. These losses, especially of nitrogen, are partly accounted for by fermentations which set the ammonia free and make the other constituents more soluble.

Preservation of Manure. — The fermentations in manure are checked by using plenty of litter to absorb the liquid manure. A small amount of gypsum may be sprinkled on the moist manure as a means of helping to fix the ammonia. The most important things to be done to preserve the manure are to use sufficient litter for absorbents and where it is necessary to store the manure some time before application to the soil, to keep it compact, moist, and protected from leaching. In the Eastern States it is a common practice to use a manure cellar under the barn. The manure is usually dropped through trap doors from above. Large doors are arranged so wagons can be backed or driven into the manure cellar for loading. Land plaster is generally used to keep down the odors. The manure is hauled out to the field in the spring or summer. This method preserves the manure fairly well, especially when it is kept compact. The chief objection to this plan is its unsanitary features.

Another provision made for protecting manure is an open shed. In some cases this is in the form of a lean-to along the side of the barn, and the manure is thrown out of the windows under this shed. A much better plan is to build a shed over a shallow pit some distance from the barn, and haul the manure into it by wheelbarrow or carrier. With this plan the urine is generally drained into the pit which contains the manure, a concrete bottom and sides preventing escape into the ground.

There is considerable question whether it pays to provide a roof for the protection of manure or not. It is believed by some authorities that if an impervious floor is provided, sloping toward the center or built in the form of a pit with sides, the rain is beneficial rather than harmful, since leaching cannot occur, and the necessary moisture to help prevent the escape of ammonia is provided. The throwing of manure out under the eaves of the barn results necessarily in the loss of much of its value by excessive leaching, and at the same time it is very objectionable in a sanitary way to have a large accumulation of manure through which cows must wade and in which flies will breed. A barnyard cannot be kept in good condition unless such accumulation of manure is avoided.

Handling Manure. — One of the most difficult problems in disposing of manure is handling the urine. As already pointed out, it is the most valuable part of the manure, but is often allowed to go to waste. There are two systems in common use for handling urine. One is an underground cistern into which the urine goes through suitable drains from the gutters. This accumulated urine is at intervals pumped into a tank wagon and distributed over the fields from a sprinkler, or a portion is pumped over the solid manure on the spreader before it is taken to the field. The other plan is to use sufficient bedding to absorb the urine and distribute it with the dung. Where sufficient bedding can be had without too much expense, this system is the most economical of labor, and therefore the most satisfactory.

The method of handling manure which has met with the greatest favor among dairymen in recent years is loading it directly from the barn into a manure spreader. The barn is arranged so the spreader may be driven through the barn and loaded from the gutters, or an overhead carrier is used, with the track extending into the yard, so arranged that the load may be dumped into the spreader.

When a load has accumulated, it is hauled to the field and

scattered. One objection to this plan is that in freezing weather the manure must be unloaded daily to prevent freezing. At times the ground is soft, so it is not desirable to drive over it with a load. It often happens that the ground where it is intended to apply the manure is not ready for the application. There is also some difference of opinion regarding the extent to which losses occur from the washing of recently spread manure when snow melts or rain falls on frozen ground. Analyses made by the Ohio Experiment Station[1] show that manure handled in the usual manner by piling in the barnyard lost over one third of its value by exposure to three months of winter and spring weather. Manure applied directly to a field from the stall showed a value of $2.96 per ton measured increase in crops grown, while manure taken from the yard after three months' exposure showed a value of $2.15, a loss in value of 27 per cent.

Taking everything into account, the plan of spreading the manure directly from the stable is to be recommended, but it may be impossible or impractical to do this at all times. The overhead track carrier is one of the conveniences of the modern barn. The advantages are that the carrier is easily loaded and moved, and it is possible to haul the manure as far from the barn as it is wished or to place the spreader on a lower level and dump into it directly.

Material for Bedding. — It is often a serious problem to supply the dairy farm with sufficient bedding. So far no satisfactory arrangement has been provided to do away with bedding of some kind.

Bedding is used primarily to keep animals clean, and the

[1] Bulletin No. 183, Ohio Experiment Station.

material that will do this most satisfactorily must be counted the best. Another purpose, and almost as important as the first, is as an absorbent for the urine. Good material for bedding should also be itself clean, which means primarily free from dust.

The following table gives data regarding the absorptive powers of common bedding material as found by trial by Doane at the Maryland Experiment Station.

Material	Water absorbed per Pound of Bedding	Lb. of Bedding required to absorb Liquid Manure from 1 Cow 16 Hours	Lb. of Bedding required to absorb for 24 Hours
Cut stover	2.5	2.8	4.0
Cut wheat straw	2.0	3.3	5.0
Uncut wheat straw	2.0	3.3	5.0
Sawdust	0.8	8.3	12.5
Shavings	2.2	3.0	4.4

As a result of these trials, Doane states that the amount of the common bedding material needed per day per cow is as follows: —

   Cut wheat straw . . . . 2.9 lb.
   Whole wheat straw . . . . . . 2.3 lb.
   Cut corn stover . . . . 3.2 lb.
   Sawdust . . . . 11.0 lb.
   Shavings . . . . 2.7 lb.

From a sanitary standpoint sawdust stands first, followed by shavings. These materials are free from the large number of bacteria and molds that often accompany straw. The bedding to be used in any particular locality will depend upon what is available and the comparative cost.

Some objections have been raised to the use of sawdust and shavings as bedding material. Sawdust makes manure so light that loss of ammonia and sometimes loss from fire fanging occurs. Applied in large quantities to the soil, it may be injurious from the effect of the acids set free, particularly tannic acid in the case of oak sawdust. The reasonable use of sawdust on most soils, however, is not injurious, and may be very beneficial.

CHAPTER XXV

COMMON AILMENTS OF CATTLE

This book makes no pretense of giving directions for the treatment of such accidents and diseases as call for the services of the competent veterinarian. A brief discussion is made of the most common ailments of cattle that the cow owner should undertake to treat without extensive experience. The discussion of tuberculosis and contagious abortion is in the nature of advice for the owner of dairy stock, and is not expected to take the place of expert advice by the veterinarian.

Instruments and Medicine Needed. — Every manager of a herd of dairy cattle should be prepared for the ordinary emergencies that are certain to come. If a competent veterinarian is not readily accessible, it is all the more important. The following instruments are the most often needed, and it is advisable to have them on hand: —

> Milk fever outfit
> 2 milk tubes of different sizes
> 3 teat plugs of different sizes
> Trocar
> Syringe
> Drenching bottle

For those with some experience, a teat bistoury, and possibly an outfit for making the tuberculin test, should be added.

A liberal amount of a good disinfectant should always be

on hand, as almost constant use will be found for it. For this purpose some of the common coal tar preparations are suitable, or crude carbolic acid, which can be prepared in a 2 per cent solution, when applied to the animal's body, or a 5 per cent solution in disinfecting other objects, such as the floor of the barn or instruments. An abundant supply of Epsom salts should also be on hand, as occasion for using them will come often. In most herds entirely too little use is made of this important medicine. A dose of 1 to 1½ pounds of salts for the grown animal should be the first treatment in nearly all cases of sickness. In every case when an animal shows loss of appetite or sickness the cause of which is not known, a physic should be given at once and the feed reduced. A second dose after three or four days is often beneficial. If the appetite of the animal has returned, the ration can again be increased to the normal.

Drenching a Cow. — The common method of administering medicine to a cow is to mix with water and give from a bottle. This is known as a "drench." When giving a drench, the head of the animal should be elevated by tying or held by an assistant. The operator stands on the left side, and grasps the nose with the thumb and fingers in the nostrils. The bottle used should be adapted for the purpose, having a long, strong neck, such as a wine bottle. The mouth of the bottle should be inserted in front of the back teeth with the mouth on the tongue as far back as the middle. If the animal coughs, the head should be at once lowered to allow the liquid to escape from the windpipe. If this is not done, the medicine may pass down into the lungs, and cause pneumonia. Unless there is some special reason for doing so, it is not customary to give over 1 to 2 quarts at a time.

Unless the herdsman is thoroughly informed regarding the treatment of cattle ailments, he will seldom have occasion to administer medicine other than Epsom or Glauber salts except under the direction of a veterinarian.

Tuberculosis. — This insidious disease is of the greatest importance to dairy cattle owners as well as to milk consumers. Since the discovery by Koch in 1882 that tuberculosis is caused by a certain bacteria, our knowledge of the subject has so broadened that were it possible to apply at once what is now known regarding the disease, it could be entirely eradicated within a few years. The well-established fact that the disease may be communicated from the cow to the human family through the milk of an affected animal makes it necessary to give this source of infection the most careful consideration.

This disease is caused by a specific agent, an organism too small to be seen by the naked eye, but easily seen under a powerful microscope. It cannot develop in an animal from methods of handling or from the surroundings but must be communicated in some way from one animal to another. It is not inherited. The germs that cause the disease escape from the affected animal with slobber from the mouth, with the dung, and in badly affected animals with the milk. The disease spreads to the healthy cow mostly from eating or drinking out of troughs that are infected from the affected animals. Hogs following tuberculosis cattle are readily affected from the manure. If a diseased animal is placed in a healthy herd, other cases are sure to follow soon. It is believed now by the best authorities that the most dangerous animal from the standpoint of human health is often the one that may appear perfectly healthy, but which passes the germs of the disease with the manure. From this source some of the germs may find their way into the milk, making it possible for the milk to be infected even when no germs come through the udder.

Tuberculosis in a cow may run its course quickly, resulting in the death of the animal, but this seldom occurs. As a rule, it progresses slowly and the animal may have it for years without any indication of ill health. All this time the animal is a menace to the health of the people. The disease may attack any part of the animal's body, but is most common, as with persons, in the lungs.

It should be thoroughly understood that it is impossible to judge from external appearances, except in extreme cases, whether the animal is affected or not. Neither can any examination of the milk that can be made be depended upon as a reliable test of the presence of the disease. Fortunately we have in the substance known as tuberculin an almost infallible agent for determining the presence of the disease even in the smallest degree. Before many years it will unquestionably be required by law that every cow supplying milk for human food be tested with tuberculin and found free from tuberculosis, as is now done by a few cities. Even leaving the question of human health out of consideration, the dairy cow owner should have his animals tested and keep his herd free from this disease for pecuniary reasons. The breeder of pure-bred stock who can guarantee his animals to be free from tuberculosis finds it a valuable recommendation. The breeder in starting a herd should by all means start with clean animals; then if every one added to the herd later be tested, it is an easy matter to keep the disease out. It is advisable, however, to have the entire herd tested once per year to make certain the disease has not gained access.

The question is often raised as to the accuracy of the tuberculin test. It has been thoroughly demonstrated that this method seldom if ever fails in the hands of a person who understands its use. Authorities on the subject state that 97 per cent of animals reacting to the test have shown lesions on diagnosis. The test is made by first taking the temperature of each animal three or four times at intervals of two hours. Tuberculin is then injected beneath the skin with a hypodermic needle. After about 9 hours the temperature reading is again taken, and repeated every two hours until three or more readings have been made. A dairyman accustomed to the use of such instruments can conduct the test himself, if he wishes, after having first assisted some competent operator to carry out the test. However, it is always safer to have the work done by a trained veterinarian, in order that no mistakes may be made and that the test will be recognized by health officials and by prospective buyers. When the testing is done by the owner himself, he should consult with the State Veterinarian or Live Stock Board that has the matter in hand for his state.

Abortion. — This term is used by cattlemen to indicate the expulsion of the fetus at any time before completion of pregnancy. Abortion may be contagious or non-contagious. The non-contagious cases may occur as the result of injury, as a fall, or from the kick of a horse, or by being crowded in a doorway. Poor feed, especially that deficient in protein or mineral matter, sudden change of feed, severe cases of indigestion, such as bloating, may also be the cause. It is also claimed that offensive odors may bring about the same result. It is also well known that ergot may cause widespread abortion in cows. This fungus is seen as black, hard, spurlike growths that protrude from the seeds of grasses at the time of ripening. Rye grass is especially subject to ergot, and it is common in blue grass, especially in low wet places.

When a single case of abortion occurs in a herd, it is to be attributed to some accidental cause. If a number occur near together, the cause is occasionally to be looked for in ergot in the feed. More often it is due to the presence of the contagious disease.

Contagious Abortion. — This disease causes more loss financially to the dairyman of the country than any other to which cattle are subject. Investigations show that this trouble is brought about by the presence of living germs in the genital organs of the cow, which bring about a condition that causes the premature expulsion of the calf. That it is contagious is shown by the spread through a herd from an infected animal that has been brought in, and by experimental inoculation. From 50 to 75 per cent of the cows in a herd usually are affected. The remainder seem to be naturally immune. The fetus is usually expelled at the sixth or seventh month. As a rule, no marked disturbance of the animal's health occurs. In most cases the usual signs of normal parturition appear, as enlargement of the udder and vulva. When the cow is already in milk, no symptoms may be noticed. After having once aborted, a certain proportion of the cows will abort a second time, usually carrying the calf a little longer than the first time, and after this seem to become immune and do not again abort. Others, as a result of abortion, become sterile or shy breeders.

The spread of the disease is probably due in most cases to infection from the male. If one infected cow is brought into a herd, she may contaminate the male, who in turn infects the other cows. It is also considered possible for the infection to gain entrance to the genital tract by close contact with infected cows, such as lying upon soiled bedding, or by the infected excretions being carried from one animal to another by the nose of a third.

Prevention and Treatment. — Every precaution should be taken to prevent the introduction of the disease into the herd. In buying an aged bull or pregnant cows, the greatest care should be exercised to make certain the disease is not prevalent in the herd from which the animals are brought. There is no means of judging from an examination of the animals whether they carry the disease or not.

If an abortion occurs, the fetus and afterbirth should be burned or buried. The aborting animal should be isolated, and the stall where she stood disinfected with a 5 per cent solution of carbolic acid. The uterus of the animal that has aborted should be washed out with two gallons or more of a disinfectant solution, such as a 2 per cent solution of creolin or a permanganate of potash solution made by dissolving a teaspoonful in 3 gallons of water. This is done by inserting one end of a piece of rubber hose into the womb, in the outer end of which is placed a funnel. The solution is poured into the funnel at about blood heat. The tail, vulva, and rear parts of the animal are also washed with the disinfectant. The washing should be repeated in two or three days, and thereafter once per week as long as any discharge appears.

It is recommended that two bulls be kept in an aborting

herd. The heifers that are bred for the first time and have no chance to be infected with the germs of abortion should be bred to a bull that is never allowed to serve a cow that has aborted. In addition to the above, all the cows should receive hypodermic injections of a 2 per cent solution of carbolic acid in doses of 25 to 50 cubic centimeters as often as every two weeks, beginning at about the fourth month of pregnancy. In case there is special reason for fearing abortion with certain animals, it is recommended that the injection of carbolic acid is increased until the pupil of the eye is seen to dilate.

The above recommendations regarding the treatment of contagious abortion is that taught by the best veterinarians. It has been followed by many breeders with good success, while others have not been able to check the progress of the disease. It is probable that in the latter cases the treatment was not thorough in some particulars. Halfway measures are of no value in a case of this kind.

Where only one herd bull is used in a herd when signs of abortion occurs, or in case the bull is a valuable animal and has been exposed to abortion, care should be taken to disinfect the sheath before and after serving any of the cows in the herd. This can best be done by irrigating the sheath with any of the solutions mentioned above by using a large syringe. It is also recommended to give the bull hypodermic injections of carbolic acid and keep him isolated from the herd.

Udder Troubles. — One of the most common troubles with dairy cows, especially among highly developed milk producers, is inflammation of the udder. It varies in severity from a mild case, when the milk is slightly stringy for a few days or a slight swelling is found in the udder, to severe cases where the udder becomes so swollen that no milk can be drawn, and which may end with the permanent loss of the udder.

Congestion of the Udder. — With heavy milkers as a rule the udder is enlarged and more or less hot and tender just after calving. This swelling may extend forward to some extent on the abdomen. This condition is to be expected, and need not cause any anxiety. It is more pronounced when the animal has been well fed and is in good flesh. When this condition exists, the animal should not receive much grain until the udder softens. The ration should be laxative in nature, and of a light character. Bran is especially adapted for feeding at this time. The milk should be drawn several times during the day, followed by active rubbing or kneading of the udder. The cow should be kept from exposure to cold weather and to cold drafts, and off cold, wet floors, until the swelling leaves the udder.

Inflammation of the Udder. — This common trouble is also known as mammitis or as garget. It varies greatly in severity. Many times the symptoms observed are swellings in the udder that do not even interfere with the milk secretion beyond causing a tenderness of the udder for a few days. Or the milk may be lumpy and full of threads, with no noticeable hardness in the udder. The milker should observe the condition of every cow carefully when milking, and report any abnormal condition noticed at once to the herdsman or take such action as seems necessary. Prompt action is always advisable, lest the conditions become severe. Such light attacks probably come from a variety of causes. It is generally believed that certain bacteria of the streptococcus group are responsible for udder troubles, but if the animal is in good condition it will resist such an attack. Any condition of the animal that lessens the power of resistance makes it possible for the trouble to start. Such a condition of the animal may be brought about by exposure to severe weather, lying with the udder on a cold floor, from injury to the udder by bruises, or by improper or too heavy grain feeding. In many cases, however, no special cause can be assigned. Mild cases, as above described, usually respond to treatment if taken in time. The grain ration should always be reduced to one third the usual amount or less at once when any inflammation appears, and kept there until the condition disappears. A physic should also be given at once, and care taken not to expose the cow to cold weather or cold drafts. An ounce of saltpeter per day for two or three days is generally beneficial after the purgative has begun to work. The cow should be milked with great gentleness, and preferably three or four times per day. If the udder is extremely sensitive, a milking tube should be used for a few days.

Severe Cases of Inflammation. — Occasionally severe attacks come on, and usually suddenly. These most often affect heavy milkers. The first symptom is a shivering of the animal, with cold ears and horns, followed in a short time by a fever. One or more quarters of the udder swell and become very hard, which is most often the first symptom seen, while the whole gland is decidedly hot and tender, and no milk can be drawn. Usually a small amount of yellowish watery fluid containing clots of casein replaces the milk. If the inflammation cannot be reduced within a short time, that quarter of the udder will not secrete any milk during that milking period, and perhaps will be permanently lost. In some cases the quarter will again secrete milk. In others a fibrous mass may develop following such an attack, or an abscess may result, which fills the udder with pus and finally discharges either through the teat or through an opening in the side of the udder.

Treatment of such severe cases must be prompt and thorough, or permanent injury, as described, will result. The cow suffers great pain from the weight of the udder. A special udder support should be on hand, or a sheet passed around the body to support the weight of the udder. Under this support next to the udder may be packed soft rags, which are kept as hot as the animal will endure by pouring on hot water every few minutes for an hour or two. At the end of this time the udder may be dried and thoroughly rubbed and kneaded for some time. At this stage an application of antiphlogistine can be made with advantage. This material is warmed until soft by placing the can in warm water. It is then applied in a layer about one fourth inch thick, with the teats protruding so the milk may be drawn. A layer of cotton is then applied over the antiphlogistine, and the udder support put in place. In about twenty-four hours the material loosens and may be removed. If the inflammation is still present, a second application should be made.

If it is impossible to apply antiphlogistine, the udder may be packed in ice, which is replenished as fast as it melts and allowed to remain several hours. In the beginning of any treatment for a severe attack of inflammation of the udder a drench should be given containing 1 to 1½ pounds of Epsom salts. One ounce of saltpeter (potassium nitrate) is also given to stimulate the action of the kidneys, and may be continued for several days.

Lice. — During the winter season especially, cattle are often affected with lice. Calves and young cattle are most often affected, but older cattle are not exempt, and when in an unthrifty condition may suffer badly from this pest. There are three kinds of lice that affect cattle. The species generally known as the blue louse, which sucks the blood, is the most common and most injurious. This species is found most numerous upon the neck and shoulders. The eggs are attached to the hair, and are known as nits. The red louse, which is less common, may be found on any part of the body, but most numerous on the neck and at the root of the tail.

The presence of lice may be suspected from the rubbing of the neck and shoulders on trees, posts, etc., and when badly infected the hair usually begins to come out in spots. Several substances may be used to kill the lice. The several coal tar dips and compounds on the market may be employed with success. The most satisfactory treatment will be found to be the use of kerosene emulsion. To make this dissolve one half pound of hard soap in one gallon of boiling soft water. As soon as the soap is dissolved, add two gallons of kerosene, mix by pumping with a spray pump or by other means until a thick creamy emulsion is formed from which the oil does not readily separate. Before use add this mixture to 19 gallons of water. The emulsion may be applied with a spray pump, or with a brush, wetting the entire animal thoroughly.

Pink Eye. — This is a contagious inflammation of the eyes, common in many herds. It usually occurs during the latter part of the summer. It is known by a discharge from the eyes, accompanied by an intense inflammation of the mucous membrane. The eyelids swell, and the eye becomes opaque. The eyes are kept shut, and the animal is often blind for several days. In some cases the animal soon recovers without injury, while in others loss of the eyesight may result if not properly treated.

The affected animal should be kept in a dark, cool stable, and supplied with easily digested food and plenty of water. The eyes should be washed twice daily at least with a strong solution of boracic acid (1 dram dissolved in 4 ounces of water). This wash should be applied directly to the eyeballs, and is conveniently done by the use of a syringe. The animal will usually recover within a few days.

Foot Rot. — This is the name applied to a common inflammation that occurs between the toes and may extend above the hoof. It is attributed to the irritation of stable manure, or some foreign substance such as a stone or cinder becoming wedged between the toes. It commonly affects sheep and cattle.

Animals running in stony lots or pastures become affected with foot rot quite often. It occurs, however, at times under conditions that leave no doubt that it is contagious. It is recognized by a limping gait and a swelling above and between the claws. The odor of the affected part is very offensive. If neglected, a serious condition may develop; but if treatment is given during the early stages, it is easily remedied. The most simple method of treatment is to clean the affected parts by means of passing a small rope between the claws and drawing it back and forth in a sawlike motion, and then applying some good disinfectant to the affected parts. Several applications of some of the coal tar disinfectants is usually sufficient to heal it up. These are best applied in their pure form. A solution of carbolic acid, 1 ounce to a pint, of water or a saturated solution of blue vitriol (copper sulfate) will also give good results. In advanced cases where the foot has swelled to any extent the application of flaxseed poultice is to be recommended.

Bloat. — This trouble comes from the formation of an excessive amount of gas in the paunch. It often results from pasturing on clover, but may occur with any kind of feeding. It is known by the excessive swelling of the left flank. If relief is not obtained in time, the animal dies from suffocation due to the great pressure on the lungs. In mild cases driving the animal at a rapid gait for some distance may be sufficient. Cold water thrown in quantities upon the cow's sides may reduce the pressure. In very severe cases the gas must be removed without delay. This is best done by the use of a trocar. In using this instrument a spot is selected equally distant from the last rib, the hip bone, and the backbone. The skin is cut for about an inch, then the trocar is thrust into the paunch. The sheath of the trocar is allowed to remain in the opening as long as any gas escapes, which may be several hours. It is generally advisable to give a dose of 1 to 1½ pounds of salts after a case of bloating.

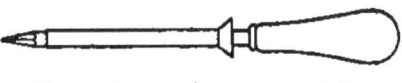

FIG. 57. — Trocar used for bloat.

INDEX

Abortion, guarding against in buying, 114; cause of, 328; treatment for, 330.
Ash in feeds, 275.
Ayrshire, Advanced Register for, 69; birth weight of calves, 174; characteristics of, 66, 68; origin of, 63; records of, 68, 70; scale of points for, 70, 72; types of, 67.

Barns, arrangement of cattle in, 305; basement type of, 296; covered yard system of, 298; floors for, 300; gutters for, 306; location of, 299; lighting of, 299; mangers for, 308; one story or shed, 297; types of, 295; round type, 298.
Bedding, material for, 321; amount required of, 322; sawdust and shaving for, 322.
Beet pulp, composition of, 278; use as feed, 291.
Bistoury or teat slitter, 225.
Bloat, 337.
Bos primigenius, 11, 12.
Bos sondaicus, 10.
Bran, composition of, 278; characteristics as a feed, 289.
Breeding registered herds, 112.
Breeds, classification of, 14; crossing of, 113; origin of, 12.
Brewers' grains, composition of, 278; use as feed, 291.
Brown Swiss Cattle, characteristics of, 76, 78; conditions in native home of, 75; origin of, 74; records of, 78, 79.
Bull, age for service, 169; age to select, 167; dehorning of, 172; feed and management of, 168-173; results at University of Missouri, 155; selection of, 154, 168; influence of age of dam upon, 168; variation in transmission of dairy qualities by, 156, 161; stalls and shed for, 170.

Calves, calf raising, 174; amount of milk to feed, 182; birth weights of, 174; cholera in, 198; feed required to raise, 179, 180; grain for, 180, 185; hay for, 186; milk substitutes for, 192; prepared meals for, 194; navel infection in, 198; supplements to skim milk, 184; temperature of milk for, 183; scours in, 197, 198; spring *vs.* fall calves, 189; ties for, 196; veal feeding for, 199; whey as feed for, 193.
Cattle, Dairy, classification of, 16; origin of, 9; marking, 211; protecting from flies, 214.
Cement floors, 304.
Channel Island breeds. *See* Jersey *and* Guernsey.
Community breeding, 109.
Corn, composition of, 278; characteristics of, 288; for growing animals, 207; stover, 288.
Cottonseed meal, composition of, 278; characteristics as feed, 290.
Covered Yard System of housing, 298.
Cow census, summary of, 109.
Cows, Dairy, age as influencing yield of milk, 148; age and per cent of fat, 151; comparison with steer in economy of production, 5; condition at calving as influencing milk secretion, 270; fat content influenced by condition at calving, 37;

care at calving time, 232; after calving, 235; drying up, 229; developing long milking period, 233; human food produced by, 6; kicking, 227; management when dry, 230; marking, 211; number of, in United States, 2; selection of, 116, 132; results from Illinois Experiment Station, 119; results from Kansas Experiment Station, 120; results from Southern States, 122; results from Storrs Experiment Station, 123; results from Iowa Experiment Station, 124; retention of after-birth in, 234; self-sucking, treatment for, 228; selection by type, 21, 132; ties for, 310; cause of variation in production, 117, 126; investigation at Missouri Experiment Station, 127.

Cow Test Association in Denmark, 146; in America, 148.

Crude fiber, 275.

Dairy farming, relation to fertility of the soil, 2; high-priced lands and, 5; advantages of, 2–8.

Dairy herd, starting a, 110; keeping free from disease, 114.

Dairy type, 17, 26.

Dehorning, advantages of, 210; dehorning bulls, 172; use of caustic potash on calves, 211.

Devons, 105.

Disease, prevention of, in starting a herd, 114.

Diseases of cattle, bloat, 337; contagious abortion, 329; congestion or inflammation of the udder, 331; treatment for, 332; drenching a cow, 325; foot rot, 336; pink eye, 335; tuberculosis, 326; garget, 331; instruments and medicines needed for, 324.

Drying up a cow, should it be done, 229; methods of, 229.

Dual-purpose cattle, definition of, 87; adaptations of, 88, 90.

Dutch Belted, characteristics of, 83; origin of, 81.

Ether extract, 276.

Feeding, amount to feed, 260; balanced rations, 267, 269; necessity for feeding as individuals, 265; overfeeding and how recognized, 263, 264; rules for amount to feed, 267; relation to live weight, 264; succulent feeds, 268; underfeeding and how detected, 261, 264.

Feeding Standard, Wolff's, 277; Armsby's, 280.

Feeding stuffs, composition of, 274, 278; fertilizing constituents of feeds, 3.

Feeds, constituents of, 274; fertilizing value of, 3; table of composition, 278.

Flies, protection from, 214; species of, on cattle, 215.

Floor, material for barn, 300.

Foot Rot, 336.

French-Canadian Cattle, 83.

Gluten Feed, composition of, 278; characteristics as feed, 290.

Granitoid for barn floor, 304.

Guernsey cattle, Advanced Register for, 59; conditions in native home of, 56; form and characteristics of, 57, 58; origin of, 55; records of, 58, 60; scale of points for, 61.

Gutters for barns, 306.

Hay, composition of various kinds, 278; characteristics of legume hay as feed, 288; for calves, 186.

Heifers, age at first bulling, 209; age to breed, 204; corn in ration for, 207; development of the dairy, 203; milking the, 218; factors influencing the size of, 208; influence of overfeeding when young, 206.

Holsteins, Advanced Register for, 35; characteristics of, 30; families of, 35; origin of, 27; records of, 33, 38; scale of points for, 38, 39; seven-day records, 37.

Jersey Cattle, characteristics of, 46, 49; conditions in native home of, 43; families of, 52; importations of, 45; origin of, 42; records, 49, 53; Register of Merit for, 51, 52; scale of points for, 54; registration on Jersey Island, 44; types of, 47, 48.
Jerseys, Polled, 84.

Kerry Cattle, 85.

Lice on cattle, 335.
Linseed meal, composition, 278; characteristics as feed, 290.

Maintenance ration, Armsby's Standard, 280.
Mangers for barns, 308.
Manure, amount voided, 4, 316; composition from different animals, 317; distribution of fertilizing constituents in, 317; handling of, 320; preservation of, 319; value of, 4, 316.
Marking systems for cattle, 211.
Milk, cause of secretion, 130; bitter, 227; bloody, 226; fat content influenced by condition of cow, 37; importance of rich, 133; skim milk, composition of, 178; yield and richness as influenced by age of cow, 148.
Milk fever, air treatment for, 236; apparatus for treating, 237; how recognized, 236.
Milking, before calving, 233; influence of period between milkings, 221; Hegelund manipulation in, 222; hard-milking cows, treatment for, 224; methods of, 219; the heifer, 218.
Milking machine, 221.
Milk sheets, 138.
Milk solids, amount produced by cow in year, 6, 7; per cent in different breeds, 33, 49, 58, 68, 95, 101.
Milk tubes, 221, 225.
Milk veins, 23.

Milk wells, 23.
Molasses feeds, use of, 291.

Nitrogen free extract, 275.

Oat and oat products, composition of, 278; characteristics as a feed, 289.

Pasture, turning on in spring, 254; feeding grain on, 255; provision for periods of short, 257.
Pink eye, 335.
Polled Durham Cattle, 104.
Polled Jersey Cattle, 84.
Protein, 275.

Rations, calculation of, 274, 280; palatability of, 284; order of feeding, 284; for high-producing cows, 285.
Rations, suitable for feeding, 271; calculation of, 280.
Records, summary of Experiment Station records, 108; form for daily, 138; for yearly, 143; methods of keeping, 136–146.
Red Polled Cattle, 97; characteristics, 100; origin of, 98; records for, 101, 102; score card for, 102.

Salt, amount needed, 247; necessity for, 246.
Sampling milk for testing, 140.
Score Card, 26; Ayrshire, 70; Guernsey, 61; Holstein, 38; Jersey, 54; Red Polled, 102.
Selection of cows, 116; by type, 21; importance of rich milk in, 133; by test, 133.
Self-sucking cows, 228.
Shorthorns, characteristics of, 94; dairy records of, 95; in England, 96; origin of, 92.
Silage, composition of, 278; for summer feeding, 259; value of, in ration, 268; characteristics as a feed, 288.
Soiling system, advantages of, 248; crops for soiling, 252; increased

returns from, 250; objections to the soiling, 252.

Teats, chapped, 226; stoppage of, 221; leaking, 226.
Teat slitter or bistoury, 225.
Test Associations, 146.
Timothy hay, composition of, 278; characteristics as a feed, 287.
Tuberculin Test, 388, 114.
Tuberculosis, 326.

Udder, 24; defective udders, 25; good types of, 25; importance of, 24.
Udder troubles, garget or inflammation, 332; bloody milk, 226; chapped teats, 226; instruments for treating, 225, 324; leaking teats, 226; warts on teats, 227.

Veal, 199; age requirements for, 200; feeding for, 200; production in Europe, 202.
Ventilation, 311; forces causing ventilation, 312; King System, 312; necessity for ventilation, 311; window, 315.

Warts, on teats, 227.
Water, amount required for cows, 241; warming water for, 244.

Fig. 49. — A good example of a round barn. Built at the University of Illinois.

Fig. 50. — Interior arrangement of a good barn.

Fig. 48.— A one-story or shed type barn. At the Kansas Agricultural College.

Jersey heifer, light fed, calved at 23 months, age 3 years, weight 622 pounds, height at withers 46.1 inches.

Jersey heifer, heavy fed, late calving, calved at 37 months, age 3 years, weight 1194 pounds, height at withers 53.1 inches.

FIG. 41. — Light fed early calving compared with heavy fed late calving.

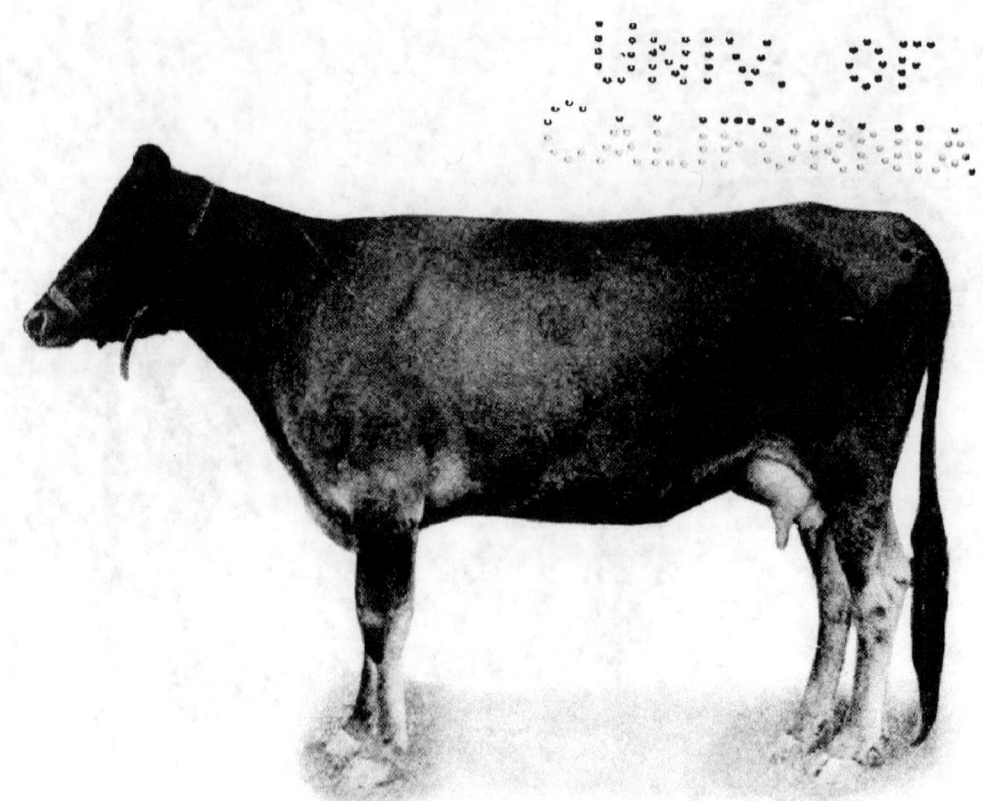

Jersey heifer, light fed, age 4 years, calved at 22 months, weight 701 pounds, height at withers 47 inches.

Jersey heifer, light fed, age 4 years, calved at 35 months, weight 816 pounds, height at withers 49.1 inches.

FIG. 40. — Influence of age at first calving upon size.

Holstein heifer, light fed, age 1 year, weight 396 pounds, height at withers 42.3 inches.

Holstein heifer, heavy fed, age 1 year, weight 730 pounds, height at withers 48.2 inches.

FIG. 39. — Influence of feed upon size and conformation.

Fig. 36.— Bull shed built at Purdue University.

Fig. 29. — "Bessie Bates," a pure-bred Jersey; bred by the University of Missouri. Record, 13,885 pounds milk, 680 pounds fat. A good type and a great producer.

Fig. 30. — An example of the difficulty in selecting by type. Of fair dairy type, but a very inferior dairy cow.

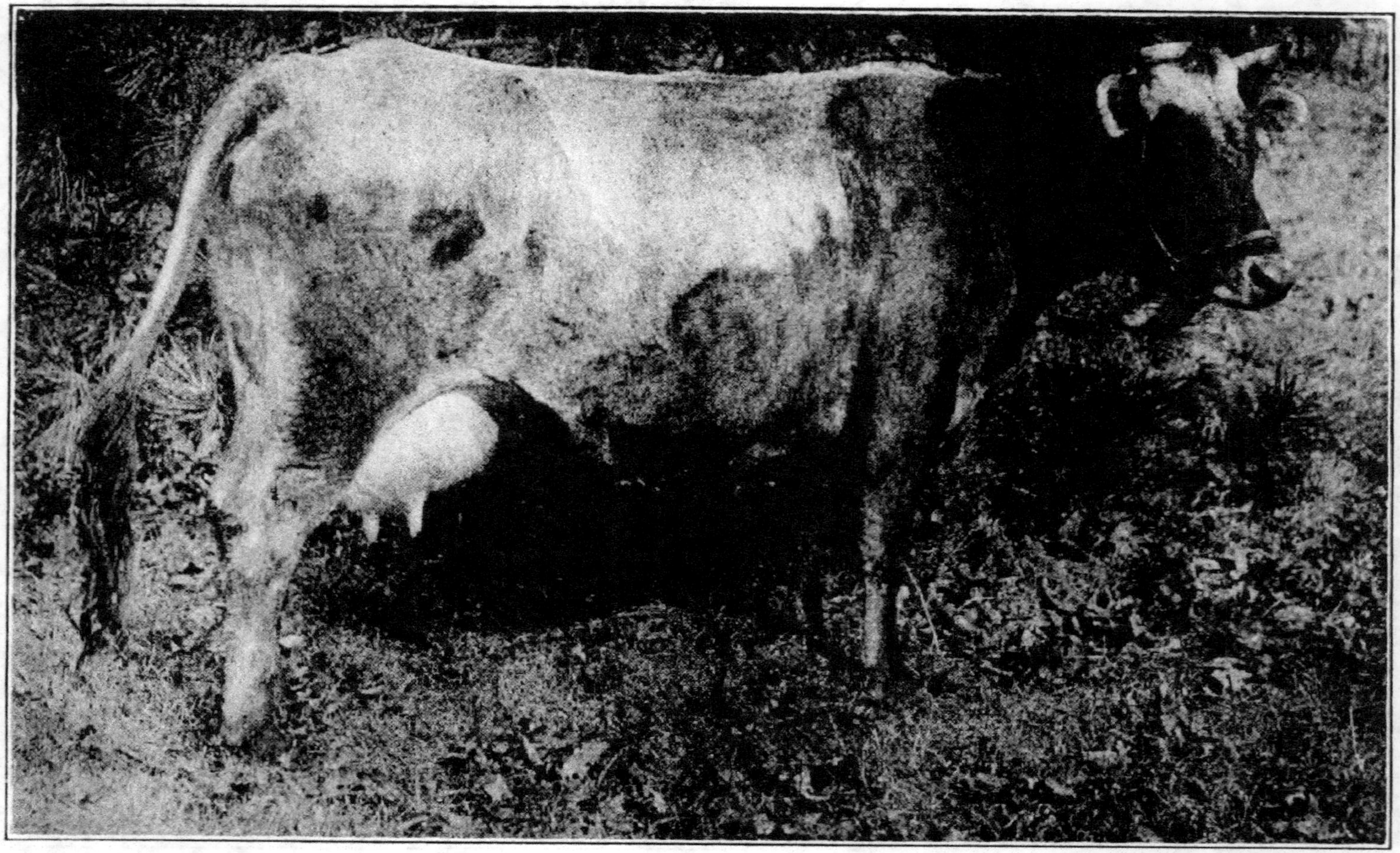

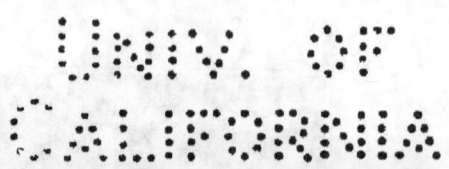

FIG. 28.—Jersey cow, "Pedro's Ramaposa." The better cow in the experiments reported. Figure 6 is the inferior cow.

Fig. 27. — Red Polled cow. A good representative of the dairy type of the breed.

Fig. 25. — Pure-bred Shorthorn cow, "Lula." Dairy type. Record for year, 12,341 pounds milk, 515 pounds fat.

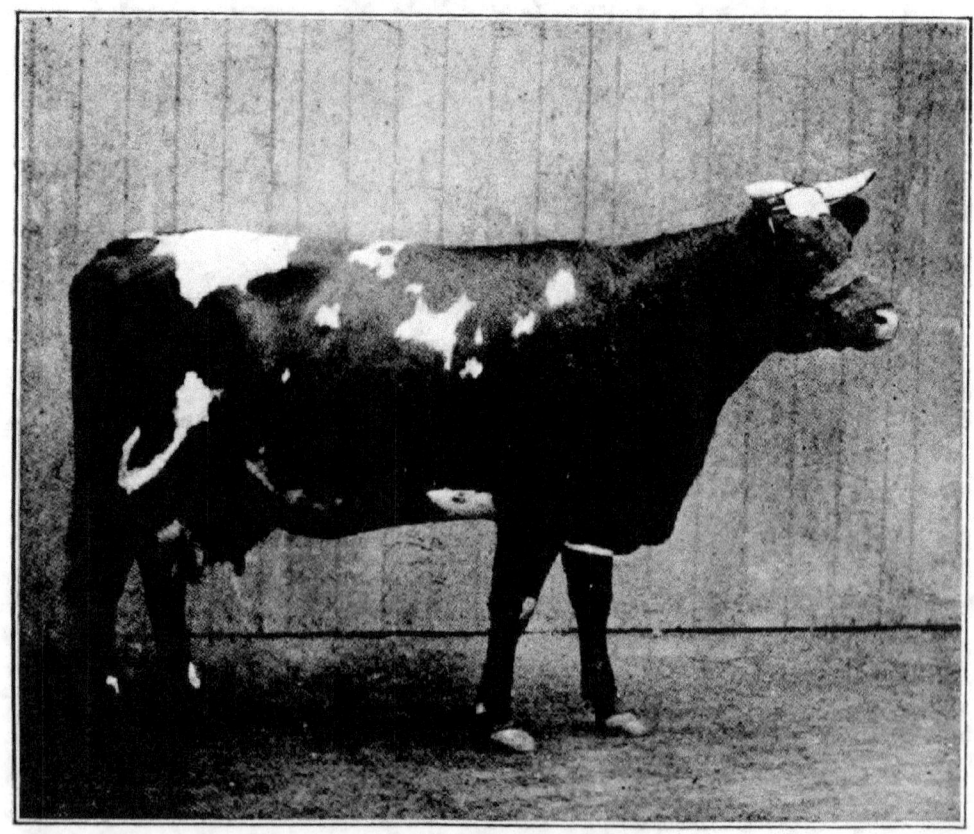

Fig. 26. — Pure-bred Shorthorn cow, "Lady Stratford." First prize, London Dairy Show.

FIG. 24.—Group of Dutch Belted cattle, as exhibited at the Illinois State Fair.

Fig. 23. — Brown Swiss cow, "Onetta." Winner of many show ring prizes.

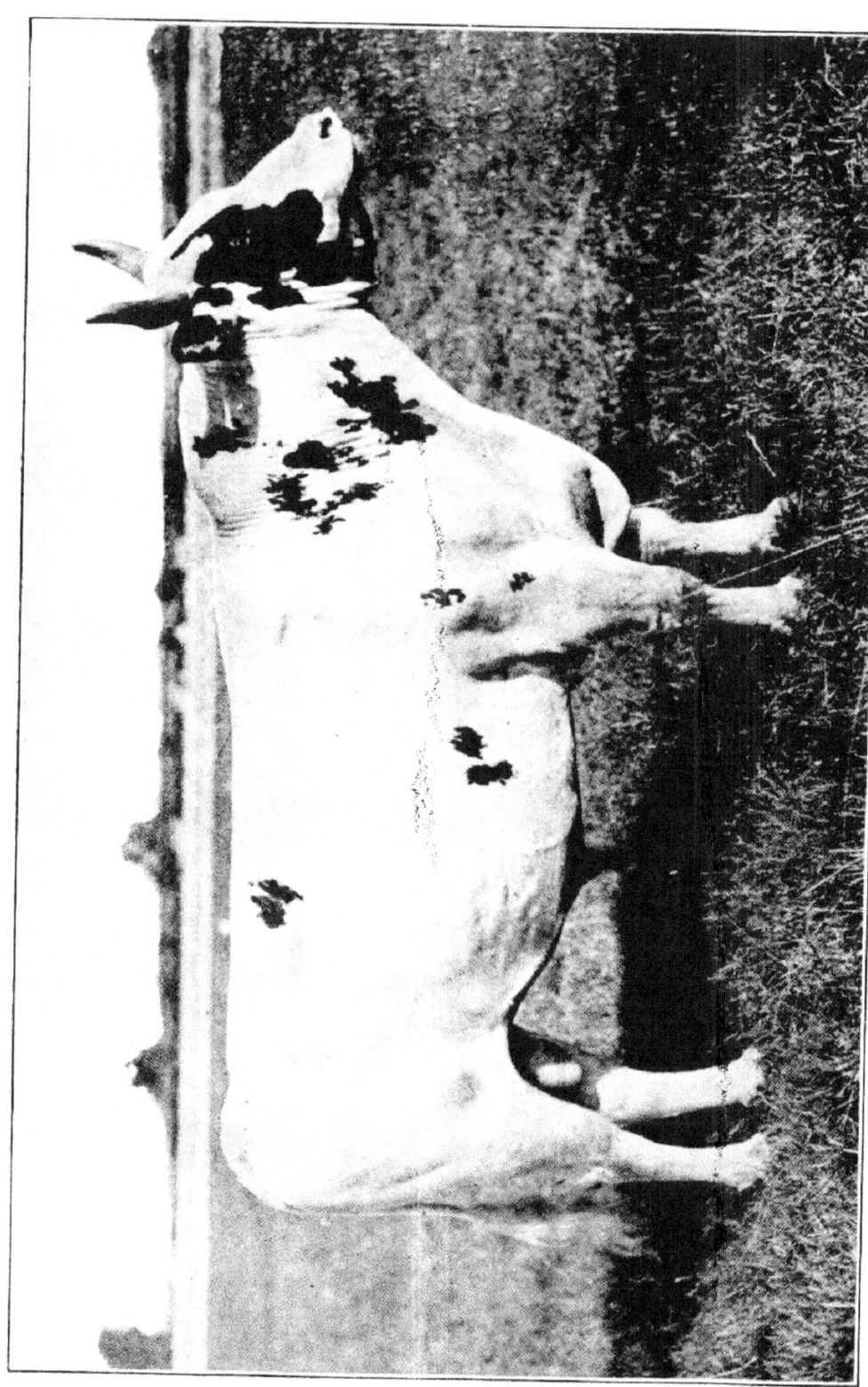

FIG. 22. — Ayrshire bull. A famous prize winner.

Fig. 21. — Ring of aged Ayrshire cows at Alaska, Yukon Exposition. A remarkable group.

Fig. 20. — Guernsey bull, "Imported King of the May." Sire of 6 daughters in Advanced Register.

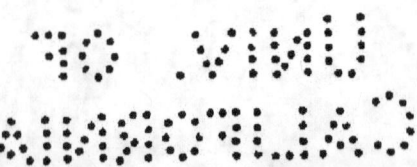

Fig. 19. — Pure-bred Guernsey cow, "Dolly Dimple." A remarkable producer. Record, 18,458 pounds milk, 907 pounds fat in one year as a four-year-old.

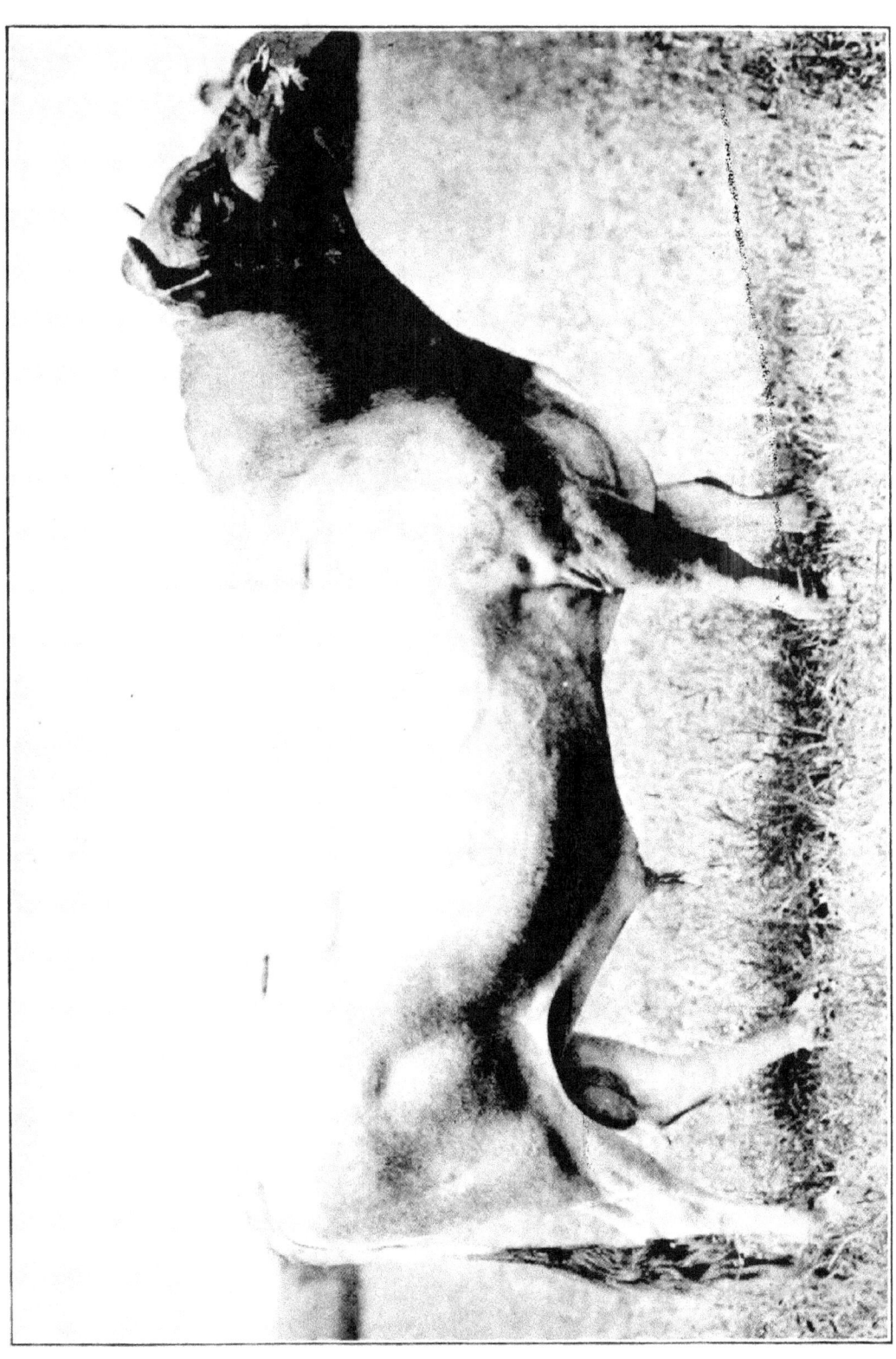

FIG. 16. — Pure-bred Jersey bull, "Hood Farm Pogis 9th." Sire of 23 daughters in the Register of Merit.

Fig. 17.—Group of American type Jersey cows bred by the University of Missouri. Average record, 12,440 pounds milk, 602 pounds fat in one year.

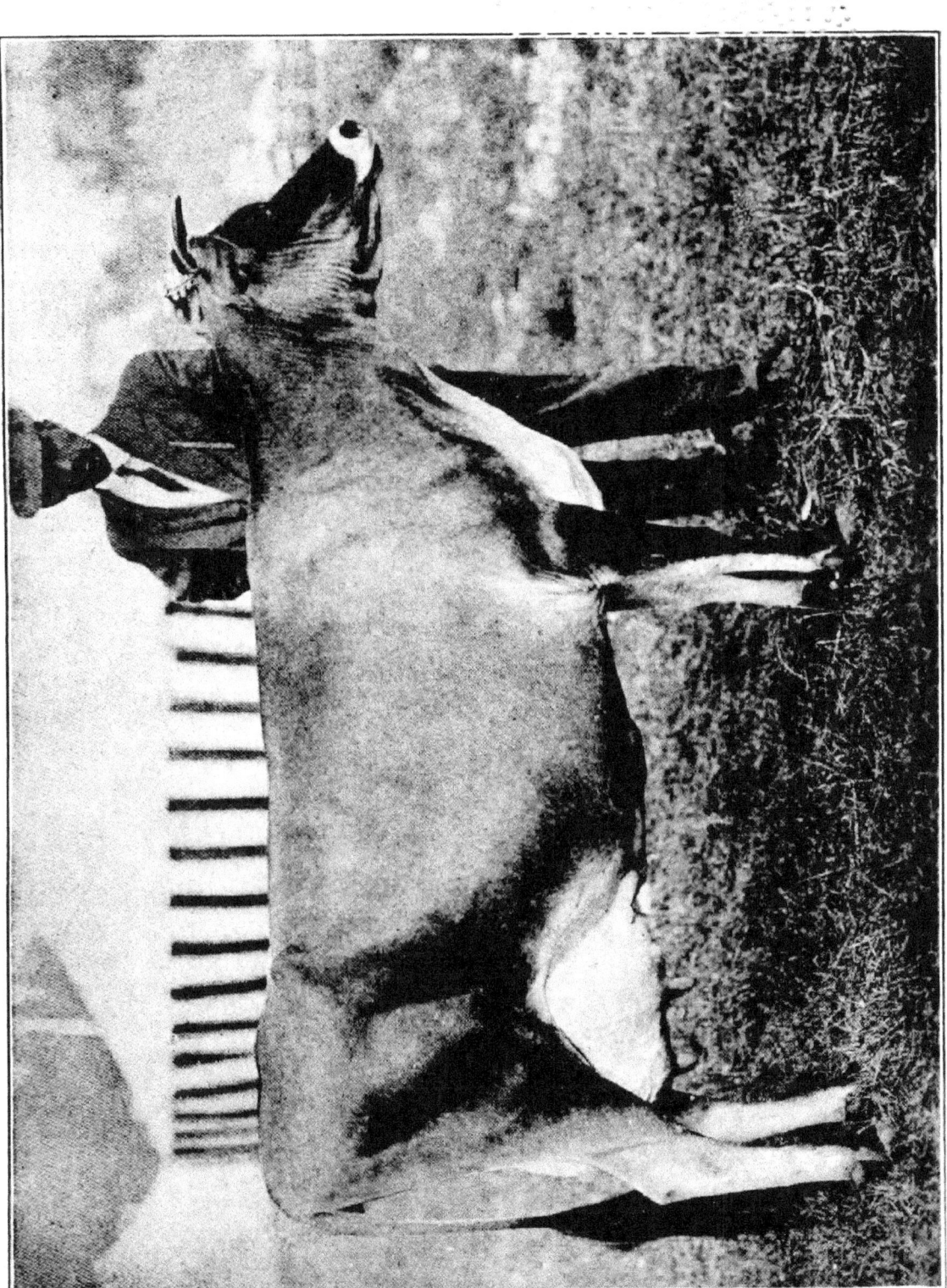

Fig. 16. — Pure-bred Jersey cow, "Bosnian's Anna." Good example of the island type. A famous show cow and a great producer.

Fig. 15. — Holstein bull, "Hengerveld DeKol," sire of 114 daughters in the Advanced Registry. One of the most remarkable sires known among dairy breeds.

"Pontiac Gerben DeKol." A good type Holstein heifer. Record as 4-year-old, 17,692 pounds milk, 519 pounds fat.

FIG. 14. — Pure-bred Holstein cow, "Missouri Chief Josephine." Record, 28,861 pounds milk, 741 pounds butter fat in one year.

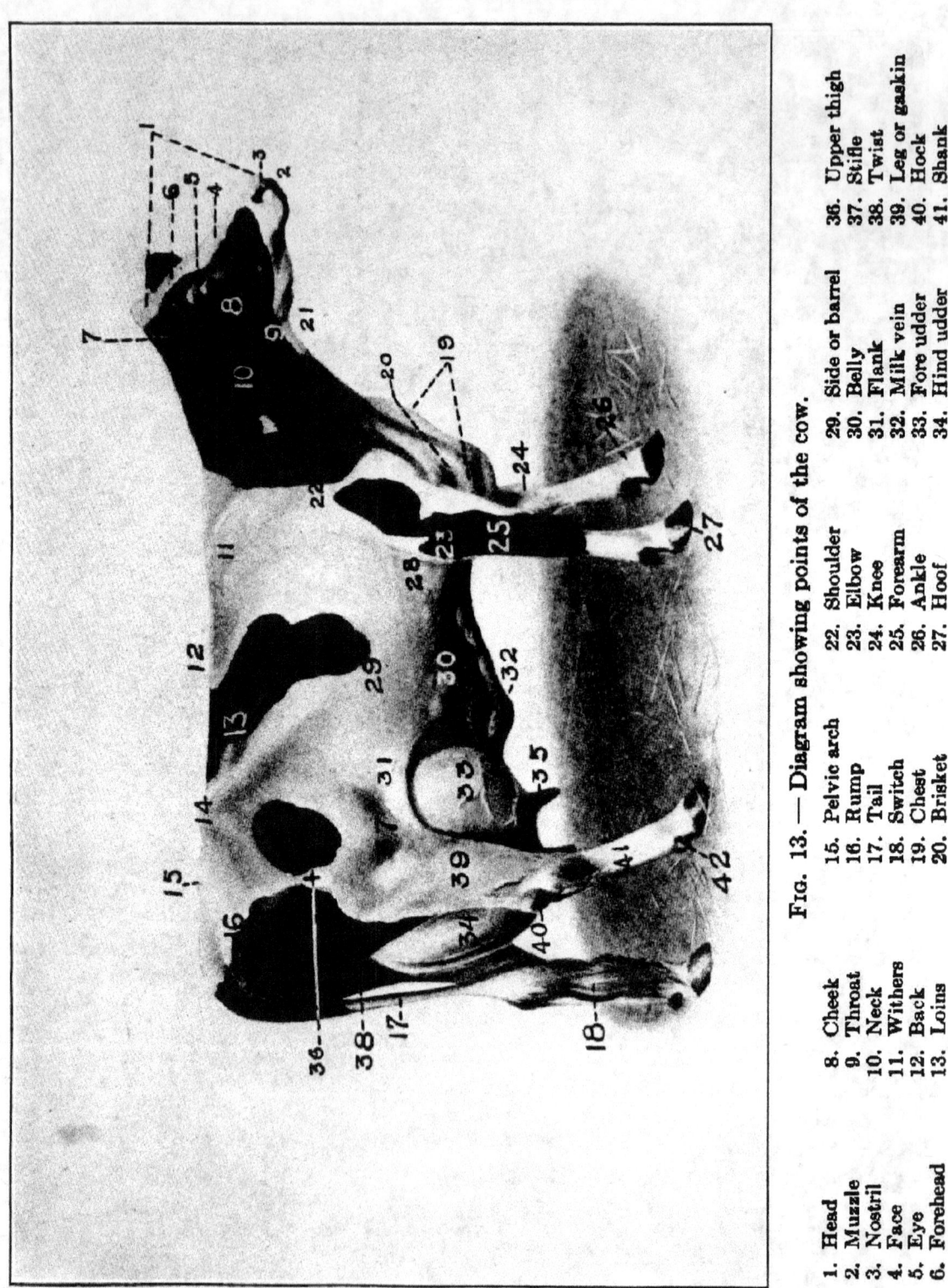

Fig. 13.—Diagram showing points of the cow.

1. Head	15. Pelvic arch	29. Side or barrel
2. Muzzle	16. Rump	30. Belly
3. Nostril	17. Tail	31. Flank
4. Face	18. Switch	32. Milk vein
5. Eye	19. Chest	33. Fore udder
6. Forehead	20. Brisket	34. Hind udder
7. Ear	21. Dewlap	35. Teats
8. Cheek	22. Shoulder	36. Upper thigh
9. Throat	23. Elbow	37. Stifle
10. Neck	24. Knee	38. Twist
11. Withers	25. Forearm	39. Leg or gaskin
12. Back	26. Ankle	40. Hock
13. Loins	27. Hoof	41. Shank
14. Hip bone	28. Heart girth	42. Dew claw

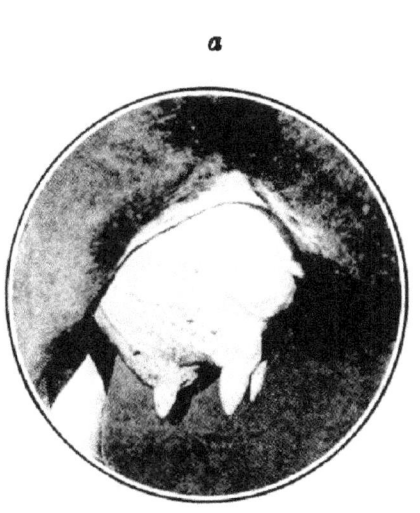
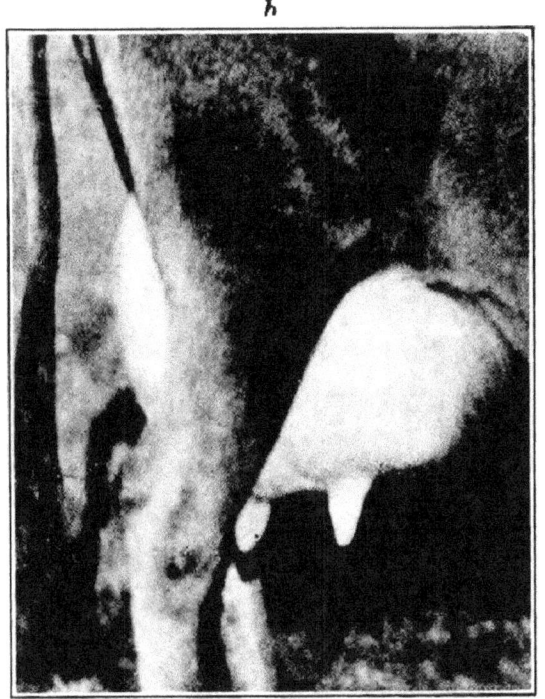
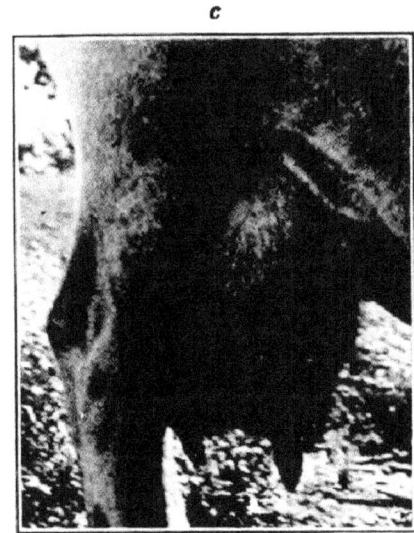

Fig. 12. — Defective udders. *a*, weak fore quarters; *b*, a fleshy udder; *c*, a pendulous udder; *d*, udder greatly lacking in capacity, especially in front.

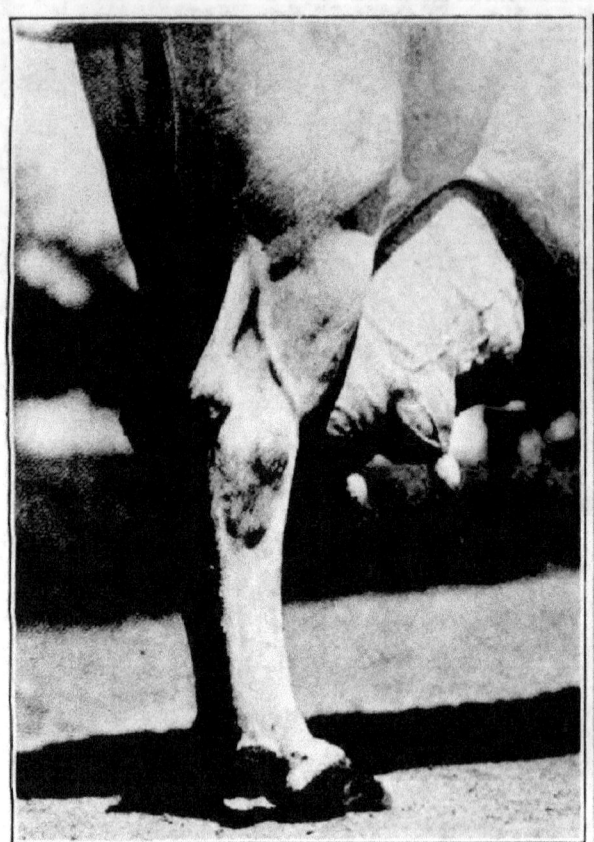
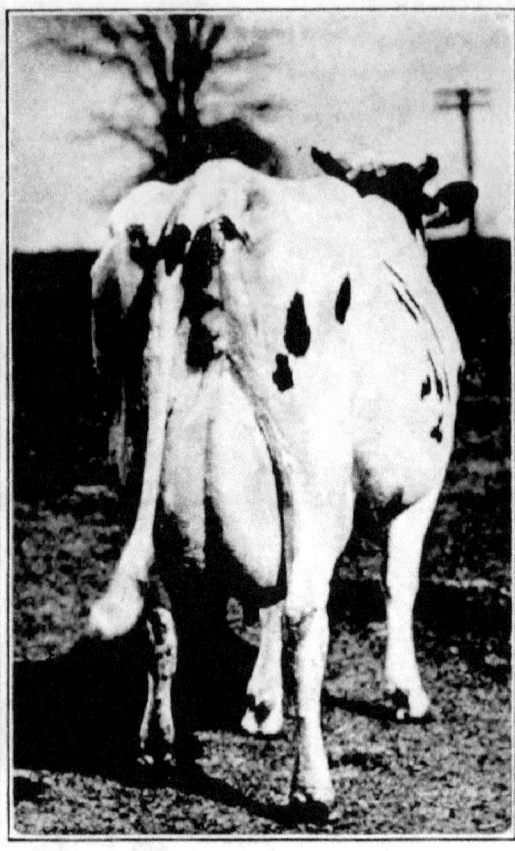
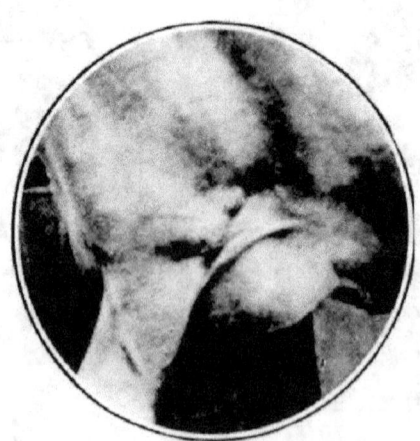

Fig. 11. — Examples of well-formed udders.

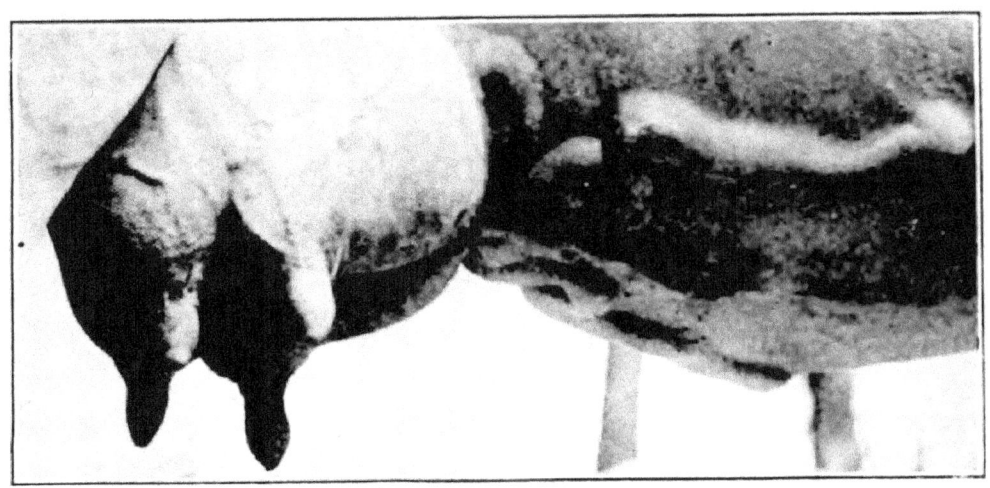

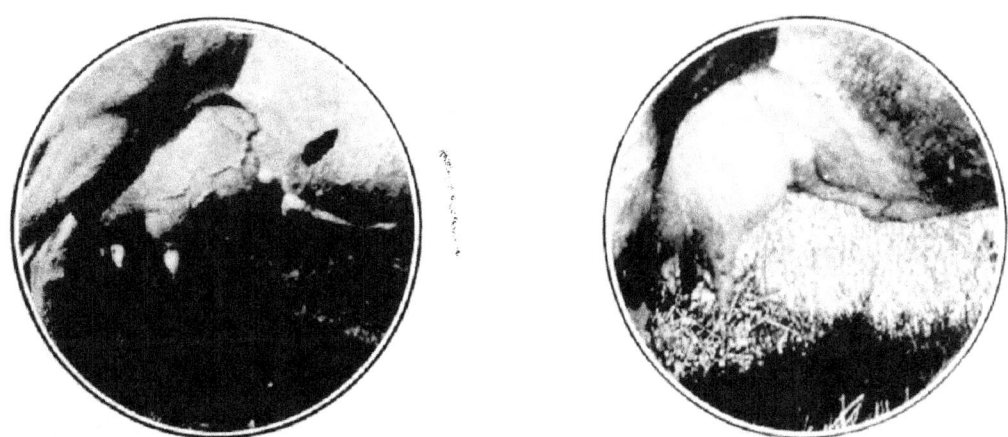

Fig. 10. — Examples of well-developed milk veins.

Fig. 9 — "Pedro's Estella." A pure-bred Jersey cow of great capacity. Record as a 3-year old, 11,063 pounds milk, 605 pounds fat.

Fig. 8. — Pure-bred Ayrshire cow, "Auchenbrain White Beauty 2d." A good type of a dairy cow.

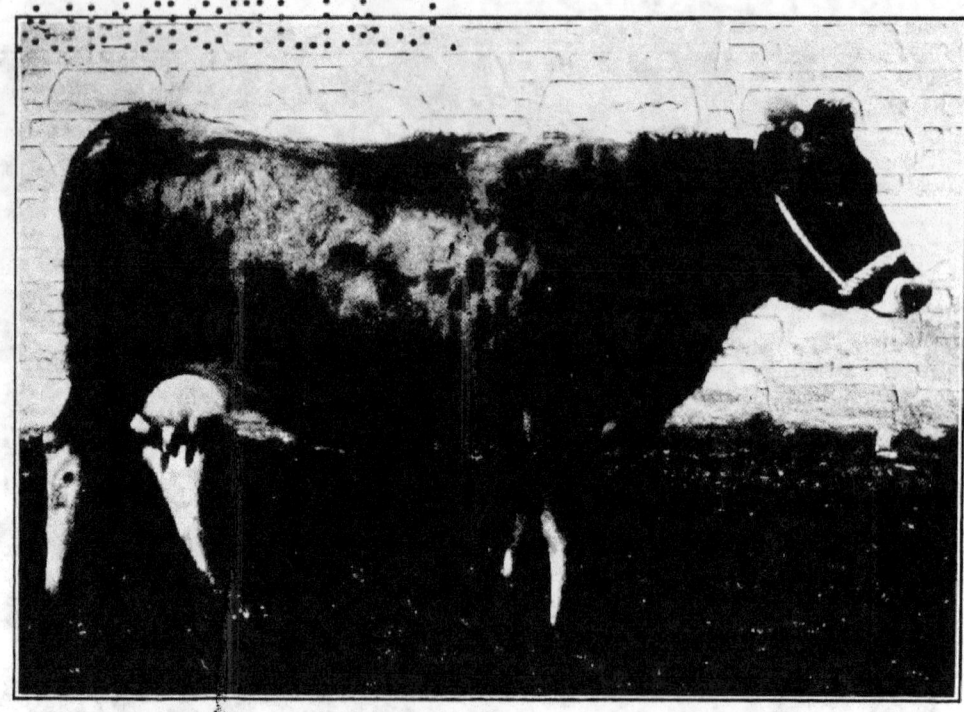

Fig. 6. — Pure-bred Jersey cow. Lacks dairy temperament. Average record for four years, 2501 pounds milk, 122 pounds fat.

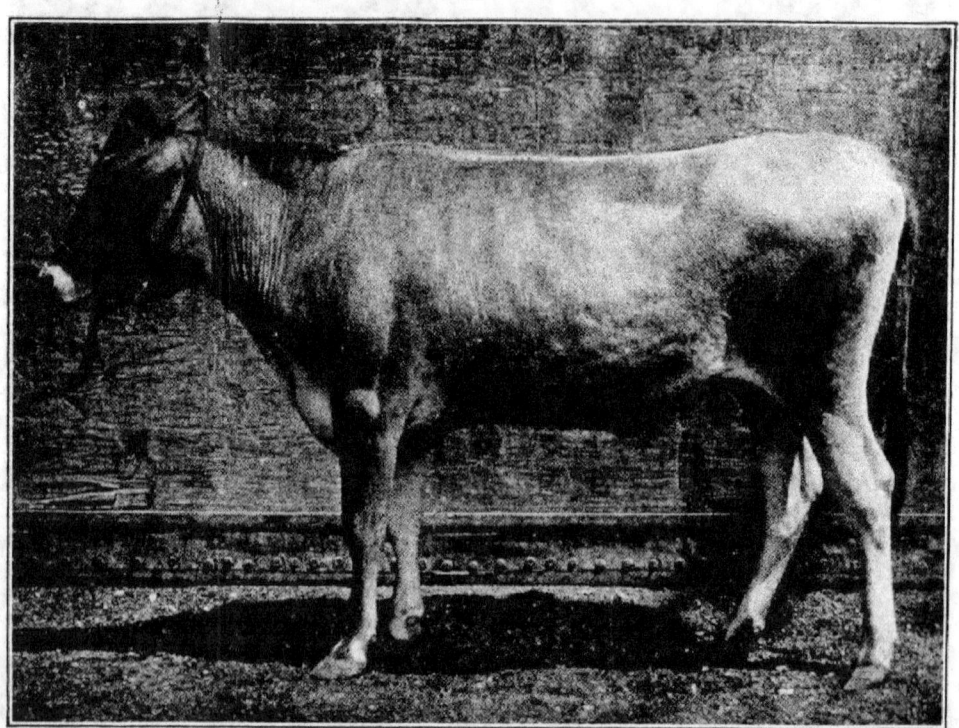

Fig. 7. — Pure-bred Jersey cow. Lacks capacity and dairy temperament. Record for one year as 2-year old, 1260 pounds milk, 62 pounds fat.

Frontispiece

FIG. 1.— Group of dairy cows exhibited by University of Missouri at the Missouri State Fair. Six Jersey cows with average record of 12,440 pounds milk and 602 pounds fat; one dairy Shorthorn, record 12,341 pounds milk, 515 pounds butter fat; two Holsteins, average milk record 22,633 pounds, butter fat record 680 pounds.

www.ingramcontent.com/pod-product-compliance
Lightning Source LLC
Chambersburg PA
CBHW062347220526
45472CB00008B/1729